INTRODUCTION

There is no such thing as Science. The word Science refers to a reified generality that together with others, like Nature and Culture, has been a constant source of false problems: are controversies in Science decided by Nature or Culture? Avoiding badly posed problems requires that we replace Science with a *population of individual scientific fields*, each with its own concepts, statements, significant problems, taxonomic and explanatory schemas.[1] There are, of course, interactions between fields, and exchanges of cognitive content between them, but that does not mean that they can be fused into a totality in which everything is inextricably related. There is not even a discernible convergence towards a grand synthesis to give us hope that even if the population of fields is highly heterogeneous today, it will one day converge into a unified field. On the contrary, the historical record shows a population progressively differentiating into many subfields, by specialization or hybridization, yielding an overall divergent movement.

This book is an attempt at creating *a model* of a scientific field capable of accommodating the variation and differentiation evident in the history of scientific practice. This model can only be applied to concrete fields, so every aspect of it will be illustrated with examples from the history of chemistry. This particular field has all the necessary characteristics to serve as an exemplary case: it has undergone splittings through specialization (inorganic versus organic chemistry) as well as giving birth to hybrids with other fields (physical chemistry). The model is made of three components: a *domain of phenomena*, a *community of practitioners*, and a set of *instruments and techniques* connecting the community to the domain. The domain of a scientific field consists of a set of objective phenomena.[2] The term "objective phenomenon" refers to an effect that can emerge spontaneously or that, on the contrary, might require active interventions by an experimenter to refine it and stabilize it. The former case is illustrated by the celestial phenomena studied by astronomers, while the latter is exemplified by laboratory phenomena.

The domain of any actual field will typically contain phenomena that exist between the two extremes of the given and the fabricated, the combinations being so varied that few general statements can be made about all domains. One generalization is that the contents of a domain must be *publicly recognizable, recurrent, and noteworthy*.[3]

The domain of chemistry is composed of substances and chemical reactions. A good example is the reaction of an acid and an alkali, and their transformation into a neutral salt. The chemical reaction itself had been publicly recognized to exist for centuries before chemistry became a field. The powerful effervescence produced when acidic and alkaline substances come into contact, suggesting an internal struggle or even a battle, had been considered noteworthy since ancient times. But once chemistry came into being, additional phenomena began to accumulate around this one, enriching the content of the domain. One was the effect produced when acids or alkalis interacted with vegetable dyes, acids changing them to blue, while alkalis changed them to red. This effect began as a phenomenon but it was rapidly harnessed as a tool, a reliable indicator of the acidity or alkalinity of a substance. By the middle of the eighteenth century, the products of the chemical transformation, neutral salts, had proliferated and become the most important member of the domain: the chemist had learned to synthesize a neutral salt not only by the reaction of acids with alkalis, but also by reacting acids with metals and other bases. From those early beginnings, the chemical domain has evolved into a highly complex entity. By one calculation, the domain included over 16 million substances at the close of the millennium, with as many as a million new substances added to it every year.[4] Thus, unlike the concept of Nature, which suggests a fixed object of study in which everything is given, a domain is a growing and changing target, never ceasing to pose problems to practitioners, and constantly eluding the goal of a final and definitive account.

The second component of a field is a community of practitioners whose personal practices are shaped by a variety of *cognitive tools*: the concepts they use to refer to phenomena and their properties; the set of statements they accept as true; the taxonomies they use to give order to the domain; the significant problems on which they are working at any one time; and the explanatory strategies they use to search for solutions to those problems.[5] These various cognitive tools are what is produced by the community of practitioners, but it is also what guides and gives regularity to their daily activities. The term "tool" is used deliberately to suggest that concepts, statements, problems, explanatory and taxonomic schemas must be deployed *skillfully* to have a real effect on a field. The role of manual skills in the laboratory has been greatly emphasized in the last few decades, as has their mode of transmission: skills are taught by example and learned by doing. But skills are not the monopoly of laboratories. Any cognitive

PHILOSOPHICAL CHEMISTRY

Logic of Sense, Gilles Deleuze

Nietzsche and Philosophy, Gilles Deleuze

Proust and Signs, Gilles Deleuze

Anti-Oedipus, Gilles Deleuze and Félix Guattari

A Thousand Plateaus, Gilles Deleuze and Félix Guattari

Barbarism, Michel Henry

From Communism to Capitalism, Michel Henry

Seeing the Invisible, Michel Henry

Future Christ, François Laruelle

Philosophies of Difference, François Laruelle

Essay on Transcendental Philosophy, Salomon Maimon

After Finitude, Quentin Meillassoux

Time for Revolution, Antonio Negri

Althusser's Lesson, Jacques Rancière

Chronicles of Consensual Times, Jacques Rancière

The Politics of Aesthetics, Jacques Rancière

Of Habit, Félix Ravaisson

The Five Senses, Michel Serres

Statues, Michel Serres

Times of Crisis, Michel Serres

Art and Fear, Paul Virilio

Negative Horizon, Paul Virilio

PHILOSOPHICAL CHEMISTRY

Genealogy of a Scientific Field

Manuel DeLanda

Bloomsbury Academic
An imprint of Bloomsbury Publishing Plc

B L O O M S B U R Y

LONDON • NEW DELHI • NEW YORK • SYDNEY

Bloomsbury Academic

An imprint of Bloomsbury Publishing Plc

50 Bedford Square
London
WC1B 3DP
UK

1385 Broadway
New York
NY 10018
USA

www.bloomsbury.com

British Library Cataloguing-in-Publication Data
A catalogue record for this book is available from the British Library.

ISBN: HB: 978-1-47259-183-8
ePDF: 978-1-47259-185-2
ePub: 978-1-47259-184-5

Library of Congress Cataloging-in-Publication Data
A catalog record for this book is available from the Library of Congress.

Typeset by Fakenham Prepress Solutions, Fakenham, Norfolk NR21 8NN
Printed and bound in Great Britain

CONTENTS

tool must be applied using abilities that are also acquired by training. In addition to this, the set of cognitive tools available to a community at any one point in time will be modeled as forming a heterogeneous collection of individual items not a monolithic theory. An apparent exception to this is cognitive content that has been given an axiomatic form, transformed into a set of statements (axioms), the truth of which is beyond doubt, from which many more statements (theorems) can be mechanically derived. But far from constituting an exception, concrete axiomatizations should be considered an additional cognitive tool added to the rest, rather than the final polished form that all cognitive content should take.

The third component of a field is the instruments and procedures that act as an interface between a community and a domain. Sometimes instruments are developed by practitioners as part of a well-defined line of research. In this case, they play the role of mere tools, increasing the accuracy of measurements or reducing the noise in the information that is extracted from a phenomenon.[6] But often instrumentation plays a larger role, that of *enabling experiments* that would not be possible to perform otherwise.[7] A good example is the electrical battery (the Volta pile), an experimental device used by physicists to produce phenomena related to electricity, but that became a powerful analytical instrument in the hands of chemists. The continuous electrical current created by the battery, when transmitted through a liquid solution in which certain chemical reactions were taking place, allowed chemists to disintegrate even the most stubborn compound substances, greatly increasing the power of chemical analysis and creating an entire subfield with its own phenomena crying out for explanation: electrochemistry.

Although these three components would suffice to model a field and follow its changes through time, a fourth one must be added for the model to be complete. This is the component that replaces the reified generality Culture. Much as the members of a scientific community must be pictured as being *embodied*, possessing the necessary skills to deploy the available cognitive tools, so the community itself must be viewed as *socially situated*. In particular, the practitioners of a field typically work in an institutional organization—a laboratory, a university department, a learned society—and organizations possess an authority structure that must be legitimate (therefore involving social values) and must have the capacity to enforce its mandates (therefore involving practices that are non-cognitive). These institutional organizations interact not only with one another—as when laboratories attempt to replicate or falsify the findings of other laboratories—but with governmental, ecclesiastical, and industrial organizations as well. In these interactions, practitioners must be able to justify their claims to knowledge, a justification that often involves a variety of rhetorical strategies. As the organizations evolve in time, they also tend to develop myths about their origins that play an important role in the legitimation of their authority. This fourth component of the model is the least important because although social values and professional agendas do affect the *focus of research and its rate of advance*, it can be shown that their effect on the cognitive content of a field is minimal.[8] This statement goes against the grain of most of the sociological literature on Science produced in the last few decades, so it will have to be defended. The necessary arguments, however, will be postponed until the last chapter of the book, after we have examined what really matters for a model like this: the contingent development of the cognitive tools produced by successive communities of practitioners, as well as the contingent evolution of the domain and instrumentation.

The fourth component of the model impinges on another question: through their personal practices, members of a scientific community can improve previously developed cognitive tools, and invent entirely novel ones, but these achievements can be lost unless they are consolidated into a *consensus practice*.[9] Personal practices are extremely varied, but according to contemporary evidence this variation leads to disagreement mostly over the content of *frontier* research. The same evidence shows that over time, as the research ceases to be the cutting edge, not only can collective agreement be reached but the agreed upon content can become accumulated in textbooks.[10] Textbooks are notoriously unreliable when it comes to questions of the history of a field (mythologized genealogies often find a home in these teaching aids) or questions about scientific methodology (justificatory rhetoric often accompanies the discussion).

But once we get rid of myth and rhetoric, the content of textbooks can serve as an important guide to what has become collectively accepted, and what is being taught to the next generation of practitioners, at any one point in time.[11]

The strategy followed in this book is to use a series of textbooks, separated from each other by 50-year periods, to follow the historical formation of consensus. Chemical textbooks from 1750, 1800, 1850, and 1900 will be used to sample the content of consensus practice at regular intervals. But why stop at 1900? Because the cognitive tools that shape personal practices are characterized by both their properties as well as by their dispositions. Unlike properties, which if they are real they are also actual, dispositions can be real but not actual if their exercise is delayed or obstructed, so their full reality is only revealed with the passage of time. One disposition in particular, a cognitive tool's *improvability*, can only be documented once the tool has built a track record. Stopping the historical narrative in 1900 will allow us to use the track record that concepts, statements, problems, taxonomic and explanatory schemas have created in the twentieth century as an additional factor in the evaluation of their cognitive value.

The book is organized into three main chapters, each one dealing with one subfield of chemistry in the century in which it was developed. Eighteenth-century inorganic chemistry is discussed in the first chapter, followed by nineteenth-century organic chemistry in the second, and by nineteenth-century physical chemistry in the third. Each chapter has two sections, one dedicated to analyzing the cognitive tools characterizing each subfield in a deliberately impersonal tone, the other placing the cognitive tools in a historical context, describing how they governed the personal practices of the community—represented by chemists who played the role of exemplars of good practice—as well how they became part of the consensus. The fourth chapter confronts the question of the role of social conventions and values, authority relations and political alliances, in the history of a field. Positivist and constructivist skeptics alike use a famous philosophical problem, the problem of the *underdetermination* of theory choice by laboratory evidence, as their strategy to undermine claims to scientific objectivity. But the problem as traditionally stated is a false problem. The cases of underdetermination that can be found in the historical record are always local and transitory; that is, they always involve a few plausible rivals and are often resolved once novel evidence becomes available. The chapter concentrates on dissolving this false problem, but it also has some positive content. Social conventions may not play the constitutive role that positivists and constructivists claim, but they are real and their role in scientific practice must be evaluated. And similarly for questions about rhetoric and authority.

The bulk of the book is spent discussing the cognitive content of the field of chemistry, and half of this discussion is historical. This raises the question of why a book of philosophy should concern itself with a subject that professional historians handle so much better. The answer is that if a book's strategy is to eliminate reified generalities like Science, Nature, and Culture, it must replace them with *singular individuals*, that is, with historically unique entities: not only persons pursuing scientific careers, but individual communities and organizations, individual fields, domains, and cognitive tools. Once such an ontology is adopted, there is no choice but to take history seriously, since the identity of the entities that figure in explanations is entirely contingent. Moreover, given the reality of local and transitory underdetermination, paying attention to historical detail is not enough: *temporal periods long enough* to allow for the consequences of underdetermination to play themselves out must be studied. The choice to track the chemical field for 200 years was made for this reason, since analyzing scientific controversies over shorter periods of time has often led to erroneous conclusions. Following this strategy will force philosophers of science to master the existing historical literature, but this can only be a good thing for philosophy.

1 CLASSICAL CHEMISTRY

A Multiplicity of Cognitive Tools

We must begin by exploring the question of whether the different components of a scientific field can *be improved over time*: can domains increase in complexity and order? Can instruments and techniques get better at producing information about the domain? Can the personal practices of members of the community be extended and perfected? Of these three questions the easier to answer is the second, because each laboratory instrument has its own criterion of improvement: balances improve if they can be used to determine smaller weight differences; thermometers improve if they can be used to detect smaller temperature differences; microscopes improve if they can resolve smaller details. A similar point can be made about the cognitive tools governing personal practices. If what changes over time is considered to be an overarching theory, a general criterion to determine whether it is a real improvement over a rival can be hard to find. But replacing a monolithic theory with a variety of individual cognitive tools makes answering the third question easier because the improvement of each of these tools can be judged by its own local criterion: the procedures to *fix the referents* of concepts, to *establish the truth* of statements, to *assess the significance* of problems, to *judge the explanatory power* of schemas, and to *guide the extension* of classifications can all be specified, and fallible criteria for improvement can be given. The main purpose of this chapter is to illustrate this thesis by discussing concrete examples of the different cognitive tools used by eighteenth-century chemists. After that we will explore a closely related question: granting that personal practices can improve, can these gains be consolidated and passed to future generations or communities? Can improved personal practices become improved consensus practice?

Let's begin by discussing in more detail the three components of the field of classical chemistry: its domain, its instrumentation, and the cognitive tools deployed by its community of practitioners. The domain of chemistry, as we said, is constituted by *substances and their chemical reactions*.[1] The term "substance" refers to a macroscopic entity, like a gallon of pure water in a container, not to the molecules that compose it. A population of molecules has properties of its own, such as temperature or pressure, that cannot be reduced to its components, in the sense that a single molecule cannot be said to possess a given degree of temperature or pressure. Single molecules do have properties of their own, properties like kinetic energy or momentum, but only as part of a large enough population do these molecular properties become temperature or pressure.[2]

A similar point applies to the different collective states in which water can exist—steam, liquid water, and ice: single molecules cannot be said to be in a gaseous or solid state; only large enough clusters of molecules can have those states. In the eighteenth century, chemists interacted with those emergent macro-properties not with molecules, and this despite the fact that the substances in their domain were made out of molecules.

Chemical substances are sometimes given, like water, but in most cases they are produced in a laboratory. For this reason, the chemical domain is in constant growth, as substances that do not exist naturally are synthesized and added to it, leading to an exponential increase in the number of substances.[3] Substances that are manipulated in laboratories are typically defined *instrumentally*, by the kind of process needed to produce them in purified form. Separation and purification techniques were greatly improved during the eighteenth century, but some of the distinctions that were necessary to order substances according to their purity, such as the distinction between mixtures and compounds, took much longer to be clarified. Ignoring this complication for the moment, the separation and purification techniques can be described like this. Starting with a raw material from vegetable, animal, or mineral origin, physical operations like filtration or cutting are applied to separate homogenous from heterogeneous substances. Then, using operations like distillation and condensation, homogenous substances are separated into uniform mixtures and pure substances. Finally, using either chemical reactions or operations like electrolysis, pure substances are separated into compound substances, like water, and elementary substances, like oxygen and hydrogen. This results in a hierarchical ordering of the types of substances belonging to the domain:

heterogeneous mixtures, uniform mixtures, compound substances, and elementary substances.[4]

The domain of chemistry also includes chemical reactions, the transformations that substances undergo as they interact with each other. Chemical reactions were just mentioned as a tool involved in separation and purification, but by the middle of the eighteenth century they ceased to be mere tools and became phenomena. The first type of reaction that joined the domain was *displacement reactions*, a type often illustrated by a remarkable phenomenon: if we add powdered silver to a liquid solvent like nitric acid, the metal dissolves, uniting with the acid; if we then add copper, this metallic substance unites with the solvent displacing the dissolved silver which now precipitates to the bottom of the container; if we add iron next, the copper is displaced and forced to precipitate; and if we finally add zinc, it displaces the dissolved iron which accumulates at the bottom.[5] Although this phenomenon was known to exist prior to the birth of chemistry, it was only after its effects were carefully tabulated that it became clear that the disposition of metallic substances to unite with nitric acid was a matter of degree: copper had a stronger disposition than silver, iron stronger than copper, zinc stronger than iron, and so on. The dispositions of acids and alkalis to combine and form neutral salts were also tabulated, and as these relations became known and accepted, displacement reactions themselves became an object of research.[6]

The next component of a field is the instruments and techniques that allow a community of practitioners to interact with the content of their domain. The operations performed by chemical analysts to create pure substances are all associated with specific instruments. Some of them, like distillation, are of ancient origin and so are the instruments used to carry them out, like the *alembic* that chemists imported from alchemy.[7] Other operations are made possible by new instruments, as in the case of electrolysis and the electrical battery.[8] But instrumentation can also enlarge a domain by allowing the addition of new phenomena. At the start of the 1700s, air, fire, heat, and light were considered to be chemical substances, distinguished from the much better understood solid and liquid substances by the fact that they were imponderable or *incorporeal*.[9] They existed in laboratories either as wasteful byproducts of solid–liquid reactions (gases) or as instruments (fire), but to the extent that they were not objects of study they were not part of the domain. The incorporeals were puzzling because, on the one hand, they had capacities to affect—gases, if not allowed to escape from an apparatus through a little hole, caused explosions; fire transformed metals into powdery oxides and wood into ash—but on the other hand, they could not be weighed, or even contained in a vessel.

Gases were the first incorporeal to be tamed when an apparatus to capture them, the *pneumatic trough*, was invented in 1727: the gases produced in a chemical reaction were directed through the use of a bent tube into a vessel filled with water, and collected as they bubbled up on the other side of the liquid medium.[10] A parallel development, the ever increasing accuracy of balances, allowed chemists to accurately weigh these gaseous substances.[11] With the mystery of their imponderability gone, gaseous substances became part of the domain, alongside liquid acids and solid metals.

The third component of a field is the community whose practices are shaped by the use of a variety of cognitive tools, the first of which is concepts. The concepts denoted by the terms "elementary substance" and "compound substance" were implicit in the practice of chemists from the beginning of the century, becoming codified only later on. Substances considered to be elementary were formally defined in the 1780s as the *limit of chemical analysis*. Because this limit was relative to the state of analytical instrumentation, chemists would sometimes refer to them as "substances not yet decomposed."[12] The concept of a compound, on the other hand, was defined in terms of composition. The components of a substance considered to be a compound were discovered through the use of analysis and corroborated by the use of synthesis. In the early 1700s, chemists were already using the *analysis–synthesis cycle* to establish the identity of a compound substance.[13] The other member of the chemical domain, chemical reactions, were conceptualized as part of the study of the dispositions to combine exhibited by different substances. These combinatorial tendencies were made the object of thought through a concept variously denoted by the terms "elective attraction," "affinity," and "rapport." The basic idea is that dispositions to unite are *selective*, a selectivity conceptualized in the alchemical tradition in terms of the love or hatred that one substance had for another.[14] In 1700, these animistic connotations were still part of the concept: affinity was conceived in terms of similarity of composition, a conception still in terms of kinship or sympathy. But by the middle of the century, chemical reactions had been discovered that exhibited more complex forms of affinity that were not explainable by similarity, and the concept became instrumentally defined using displacement reactions as an exemplary case.[15]

The next cognitive tool is the statements that community members believe to be true at any one point in time. The simplest type is *identity statements* like "The sample in this container is pure sulphuric acid" or "This solid sample is pure gold," the truth of which was established by the correct application of separation and purification methods. Another significant category is *property-assignment statements*. Using modern units of measurement these can be illustrated with statements like "This sample of mercury weighs ten grams" or "This sample of water has a temperature of 45 degrees Celsius." The truth of statements like these was established through the use of instruments like balances and thermometers, as well as by the procedures for the appropriate use of the instruments. A third type were *disposition-assignment statements*. Thus, the tendency of substances to boil at a particular point of temperature could be stated unambiguously in the late 1700s with a declaration like "This sample of pure water boils at 100 degrees Celsius." Some dispositions, such as affinities, resisted quantitative measurement, but could nevertheless be used in the enunciation of qualitative truths, like "This sample of copper has a higher affinity for nitric acid than this sample of silver." Such descriptive statements about particular samples could, via simple induction, be generalized into statements like "All samples of pure water boil at 100 Celsius" or "All samples of copper have a higher affinity for nitric acid than all samples of silver."

After many such general statements had been generated, they could be given a graphic form, arranged in a table for example, to be used as didactic devices. Some tables, in fact, managed to convey to their users more than what was contained in the general statements that served as its raw materials, and should therefore be considered a separate cognitive tool. A good example are the *affinity tables* that began to be published in 1718. In the earliest example, the upper row of the table named 16 substances (acids, alkalis, metals), while the columns under each name listed the substances with which they had a disposition to combine. The position in the columns reflected a substance's degree of affinity: those immediately below the first row had the highest affinity while those in the bottom row had the lowest. Despite the fact that the original affinity table summarized results well known to apothecaries and metallurgists, it performed cognitive labor by making evident certain *relations* between substances, and by suggesting other combinations.[16] Additionally, like other classificatory schemes, affinity tables brought order to the rapidly expanding domain of chemistry, correcting old taxonomies, drawing finer distinctions between old categories, and sharpening the boundaries between different chemical species.[17]

In addition to concepts, statements, and taxonomic schemas, the personal practice of chemists was shaped by *empirical problems*. The form of an empirical problem can be modeled by a question like this: Why does this substance have these properties (and dispositions), instead of some other properties (and dispositions)?

This question can be divided into three separate parts. The subject ("this substance") represents what is not being explained or what is not considered problematic: the existence of substances with a chemical identity, in this case. This constitutes the *presupposition* of the problem. The predicate ("has these properties") is the object of the explanation.

But this object needs to be further specified to indicate what aspects of it demand explanation. This is achieved by the phrase following the words "instead of." We may want to explain, for example, why a substance has these chemical properties instead of other chemical properties; or why it has these chemical properties instead of these magical properties; or why it has these properties instead of having no properties at all. The different ways of setting these *contrasting alternatives* yield different explanatory goals. The presupposition and the contrasting alternatives define what is an admissible solution to a problem.[18] In the 1700s, the presupposition was not only that substances had an enduring identity, but also that they had certain properties because of their composition. So the problems posed really had the form: Given that substances derive their physical and chemical properties from their composition, why does this substance have these properties, instead of some other properties?

These compositional problems depended for their solution on the prevailing conception of what the ultimate chemical components were. For most of the century, the basic building blocks were thought to be four or five *principles*, arrived at as a compromise between Aristotle's elements (Air, Fire, Earth, Water) and Paracelsus' principles (Sulphur, Mercury, Salt). The terms "Sulphur" or "Mercury" did not refer to the traditional substances but to ingredients that brimstone and quicksilver had in common with many other substances. The inflammability of many materials, for example, was explained by their possession of the Sulphur principle, something that ordinary sulphur had in common with many oily substances.

And similarly for other properties, like metallicity: "Why do metals have the properties of being powdery and opaque when burned to a calx instead of being solid and shiny when they are not calcinated?" A valid solution to an empirical problem like this was something like: "Because metals are composed in part of the Sulphur principle, which accounts for their metallicity, but they lose this principle when calcinated." Use of the analysis–synthesis cycle could provide evidence for this solution, because the metallicity of a calx could be restored by burning it together with charcoal, a combustible substance that could be assumed to possess the Sulphur principle.[19] Because this reversibility was achievable in practice, chemists were justified in thinking of the referent of the concept of the Sulphur principle as something that was fixed instrumentally, like that of any other substance. This, however, made the identity of the Sulphur principle *analytically contingent*, since new laboratory instruments or techniques could reveal flaws in the purification and identification procedures, destabilizing its identity.[20]

In addition to compositional problems, the domain of chemistry confronted eighteenth-century practitioners with problems concerning chemical reactions and their products. These problems were typically solved in terms of the dispositions for combination that the reacting substances, or their components, had for each other. In other words, the existence of affinities was part of the presupposition of the problem. So the general form of a problem like this was: Given that chemical reactions are causally governed by affinities, why does this chemical reaction yield these products instead of these other products? An eighteenth-century example of an empirical problem like this would be the question: Why does a reaction of alum and vegetable alkali yield potash of sulphur, instead of another compound?[21] The answer to this question in terms of affinities was framed along these lines: the components of the first reactant are aluminum oxide and sulphuric acid, and one of these two components (sulphuric acid) has a greater affinity for the second reactant (an alkali) than it has for the other component (aluminum oxide). Hence sulphuric acid will be forcefully displaced from the original compound, joining with the alkali to form a new compound (potash) as the final product.

After solving a variety of problems like these, certain explanatory strategies began to crystallize. One way of tracking the development of these *implicit reasoning patterns* is by modeling them through the use of logical schemas.[22] Because logical schemas are modern reconstructions, we should not expect to find them in codified form in textbooks of the period. Nevertheless, as subsequent chapters will show, they can offer an invaluable guide to follow the changes that patterns of reasoning about certain types of problems underwent as new phenomena were introduced into the domain. An explanatory strategy used since the early 1700s to solve compositional problems, for example, can be modeled by the *Part-to-Whole schema*:

Question
Why does substance X have properties Y and Z instead of other properties?

Answer
PW-1. Properties are explained by the Part-to-Whole relationship between a substance and its components.
PW-2. The properties of a whole are a blend of the properties of its components.[23]

PW-3. The most basic components are those that cannot be further decomposed using existing chemical operations.

PW-4. The most basic components are few in number (four, five, or six elementary principles).

PW-5. The properties of a whole are determined by the chemical nature of its components.

A different reasoning pattern emerged after many problems posed by chemical reactions had been tackled. The logical schema with which we can reconstruct the explanatory strategies used by the middle of the century may be referred to as the *Affinity schema*:

Question

Why does the reaction of substance X with substance Y have substance Z as its product, instead of other products?

Answer

A-1. Substances possess dispositions to selectively combine with other substances.

A-2. If X and Y are elementary substances then their dispositions will determine Z: if they have affinity for each other their product will be a combination, else they will not combine.

A-3. If X or Y are compound substances then the dispositions of their components will determine Z:

> A-3a. If only one of the substances is a compound, then the elementary substance's affinity for one of the compound's components will cause it to be dislodged and join with it, displacing the other component.

> A-3b. If both substances are compounds, then they will both decompose and their components will unite following their respective affinities.[24]

A-4. The affinity of a substance for another substance is constant.

A-5. Two substances have affinity for each other if they display compositional similarity: two substances composed of the Sulphur principle, for example, will tend to combine with each other.

As we argued in the Introduction, the set of cognitive tools must be conceived as an open set, not only because new tools may be added to it in the future (mathematical models, computer simulations) but because, even in the eighteenth century, the personal practices of chemists were shaped by other factors. Most of the textbooks used as reference in the following chapters, for example, contain *practical recipes* to analyze or synthesize substances, or procedures to prepare raw materials, and these recipes and procedures should be included in the set. Nevertheless, and despite its incomplete nature, the list of cognitive tools just given has enough heterogeneity to avoid a conception of chemical theory as a monolithic entity. The question we must answer now is whether the members of the list have the capacity to be perfected over time, and whether this improvement can be judged using only *local criteria*.

Let's begin with a discussion of substance concepts and of the way in which their referent is specified. In traditional philosophy the referent of a word is thought to be determined by its meaning: understanding the semantic content of a word gives us the means to establish its referent. But it can be argued that if we want to know whether a particular piece of metal is gold we do not consult a dictionary and use the definition of the term as a criterion of correct application. Instead, we take the piece of metal to an expert, an assayer, for example, who will perform a series of chemical tests on the piece of metal to determine its nature. A mixture of hydrochloric and nitric acid (so-called "aqua regia") could be poured over the metal, and if the latter dissolves in it then it can be confidently declared to be gold. Dictionary definitions may, of course, contain some useful information about gold or other pure substances, but these are typically mere stereotypes.[25] Thus, it can be argued that the referent of concepts for chemical substances is fixed by a combination of *ostension* and *expert intervention*. Prior to the eighteenth century, referents were already fixed by the purification procedures used to produce substances, not by the meaning of their names. This is why we can confidently assert that the acid substances in the domain of 1700 were sulphuric, hydrochloric, and nitric acid, even though back then chemists used the terms "spirit of sulphur," "spirit of salt," and "aqua fortis," which have entirely different meanings. What gives us confidence is our knowledge of the techniques used to produce these three acids: distillation of ferrous sulfate, sodium chloride, and potassium nitrate, respectively.[26]

Fixing referents using a particular technique, however, could be misleading if the substance in question was producible by a variety of techniques. In this case, any procedure that avoided misidentifications was an improvement in the way in which reference was established. One candidate for this was compositional identity: if two samples could be *analyzed into the same components*, they could be considered to be the referent of the same substance concept.[27] Another candidate was the development of instruments and techniques to *measure the properties and dispositions* regularly displayed by pure substances. In the 1740s, for example, chemists began using crystal shape, boiling points, and solubility (using water as solvent) as a means to identify substances.[28] Later in the century, new concepts for properties and dispositions were added to this list to help identify different types of air, properties like comparative density (specific gravity), tendency to explode, and flame color.[29] Every new added property and disposition made establishing the referent of substance concepts easier and less error prone, and therefore represented an improvement over past practice. The existence of several ways to fix a referent, on the other hand, meant that they could conflict with one another. Thus, the air obtained during the chemical reaction that transforms mercury calx back into metallic mercury was originally identified by the process that produced it, and by a description of its composition as "Air lacking the Sulphur principle." The description turned out to be incorrect, causing ambiguity and confusion for a while. But the conflict was resolved in favor of the purification procedures, which is why we credit the chemists who first isolated this "air" with its discovery—the discovery of the gas that we call "oxygen"—despite the fact that they misdescribed it.[30]

Statements describing substances in terms of their properties and dispositions are among those that a community may accept as true. Like substance concepts, descriptive statements can be improved, but we need to specify a local criterion for this improvement by giving a definition of truth. The problem is that to this day we do not have a universally agreed account of what kind of property truth is, or of what justifies a belief that a statement is true, or even of what it is that we do when we make claims to truth.[31] Philosophers break down into two groups, one that believes that truth is a property of single statements, the other believing that truth can only be ascribed to an entire set of statements, that is, to an entire theory.[32] The first group disagrees as to what the relevant property is, some believing that it involves an isomorphism between the components of the statement (its subject, verb, and object) and the components of a state of affairs. Others reject the existence of this isomorphism between parts, invoking instead descriptive conventions that connect the entire statement to a state of affairs. There is also disagreement about what justifies a true statement: for some it is the fact that the statement can be derived from a set of self-evident truths; for others it is its usefulness in practice, or the way in which it increases the overall coherence of an entire set of statements.[33]

Chemists in the 1700s did not theorize much about truth. But one thing was clear: almost all of them rejected a priori schemes and hence would have rejected the justification of true statements in terms of their derivability from self-evident truths. If pressed, they would have accepted as justification the usefulness of a statement in the context of laboratory practices. As to the question of what truth itself is, many chemists of that century would have probably agreed with a minimal definition: a given statement is true if it asserts the existence of a state of affairs and if the state of affairs does indeed exist. In other words, an eighteenth-century chemist would have probably agreed with the following: the statement "Glauber's salt is composed of vitriol and the base of sea salt" is true if Glauber's salt is composed of vitriol and the base of sea salt.[34]

We can accept this minimal definition of a true statement as long as we conceive of truth as a semantic property of *single statements*, that is, as long as we reject the concept of truth for an entire theory. The notion of correspondence is unacceptable only in the latter case, since it implies that a theory literally mirrors a portion of the world, and this creates the problem of how to verify that this reflection is faithful given that we cannot judge the relation between theory and world from a neutral point of view. Accepting truth for single statements avoids this danger because we can replace the mirroring relation with operations that can be defined instrumentally. Thus, in the previous example, the referents of the terms "Glauber's salt," "vitriol," and "the base of sea salt" can be fixed in any of the ways previously mentioned, and the relation "to be composed of" can be checked by using the analysis–synthesis cycle. In this case, truth is reduced to reference. Accepting this definition of truth, on the other hand, does not force us to deny that the statements accepted as true at any one time are related to one another, only to reject *holism*.[35] We can accept that when two statements co-occur as part of the presupposition of a problem, or as part of the body of an explanatory schema, ceasing to believe in the truth of one does have repercussions for the truth of the other. We could even agree that these repercussions can cascade in some cases through a larger number of statements that are logically related to one another. But this does not imply that all statements are inextricably related and holistically fused into an overarching theory.[36]

Given the minimal definition of truth just outlined, what could play the role of criteria of improvement? One criterion that would have been accepted in the 1700s was *generalization*. Early in the century, several statements about the union of different kinds of acids and alkalis were believed to be true. In particular, most chemists believed that the union of the these two kinds of substances yielded a different type of substance known as a *neutral salt*. Later on, chemists came to believe that neutral salts could also be formed by acids uniting with substances that were not traditionally considered alkalis. From the statements expressing these beliefs, the following one could be generated through induction: all neutral salts are composed of an acid and a base (a vegetable alkali, a metal, or an absorbent earth).[37]

This was a very important generalization, one that summarized the extensive work on solution chemistry that had introduced dozens of new neutral salts into the chemical domain, a generalization that a majority of chemists would have accepted as an improvement over all the particular statements from which it was inferred. Another way in which statements about composition could be improved was by transforming them from qualitative into *quantitative* statements, by discovering the *proportions* in which components must be combined to yield a particular compound substance. The first requirement for this change was an assumption that had been informally made by chemists throughout the century: that in a chemical reaction the same quantity of material exists at the start and at the end of the process. The analysis–synthesis cycle, in those cases in which it could be carried out, provided confirmation for this assumption. By 1789 it had become codified as the principle of the conservation of matter (or weight) in chemical reactions.[38] The second requirement was the ability to capture all the products of a reaction, including gases, so that the weight of every substance consumed as well as those produced could be carefully established using precision balances. Assuming that matter is not created or destroyed, and that all forms of it have been properly accounted for, a statement like "Carbon dioxide is composed of oxygen and carbon" could now be expressed like this: "Carbon dioxide is composed of 71.112 parts of oxygen and 28.888 parts of carbon."[39]

In retrospect, this quantitative statement is a clear improvement over the previous one because it provides additional information that is relevant to the determination of the properties of a compound. But in the late 1700s not everyone agreed that it was an improvement because of the many potential sources of error in the determination of the numbers. First of all, chemists had only limited control over the degree to which reactions were tightly confined, to make sure that no matter had entered the experiment and no matter had escaped or gone unnoticed. Second, while liquids and solids could be reliably weighted, only the volume of gases could be directly measured, their weight calculated via estimates of their density. Because of this, two experiments rarely gave the same quantitative results, the discrepancies explained by making assumptions about potential sources of error.[40] Finally, the statement of the numbers with many decimal places was partly rhetorical, because without the means to model the statistical distribution of measurement errors it was impossible to give an argument for the significance of decimal figures. When these limitations were removed, the superiority of quantitative statements was accepted by all members of the community.

The set of accepted statements could also be improved by sheer growth, even if the added statements expressed only qualitative truths. This growth was a direct product of the proliferation of new substances that increased the size and complexity of the domain. And as the domain grew, the need to maintain it in order through the use of classification schemes became more urgent. To some extent this need could be met by improving existing taxonomies in the simplest way: by *extending* them to cover new cases. Thus, while the original affinity table contained four acids, two alkalis, and nine metals, by 1783 it covered 25 acids, 15 earths, and 16 metals.[41] Although affinity tables were greatly improved by this extension, as they grew larger they revealed problems in the concept of affinity itself. Listing substances in columns by their decreasing capacity to combine with the substance heading the column, suggested that these relations were fixed. But laboratory experiments showed that the application of heat, for example, could change the order of affinities. To accommodate these added causal factors, the table was subdivided into two parts using an old distinction: that between the "dry method" of analysis used by assayers on metals, and the "wet method" used to analyze mineral waters.[42] This subdivision which incorporated the effects of heat can also be considered an improvement.

Not all the empirical problems raised by the affinity approach to chemical reactions could be solved by making changes to the tables. The problems posed by chemical transformations also had to be improved. We said above that such a problem could be expressed by a question like: Given that chemical reactions are causally governed by affinities, why does this chemical reaction yield these products instead of these other products?

The way in which a problem is posed can be made better by adding to its presupposition further causal factors that bear on its solution, or by changing its contrasting alternatives to make them more specific. In this case there is an improvement because more relevant factors are included or because the object of explanation is contrasted with alternatives that are more relevant. We just mentioned heat (or temperature) as a factor that became increasingly important in mid-century. Then, in the 1790s a new causal factor was added: *degree of concentration*. It was noticed that the capacity of a substance to combine with another decreased in proportion to the degree of combination already reached. Concentration could, in some cases, *reverse* the direction of a chemical reaction. Thus, affinity tables predicted that sodium carbonate (soda) transformed spontaneously into common salt when reacted with calcium chloride. But if salt and calcium carbonate were present in large enough quantities, then soda spontaneously formed.[43] Improving the way in which the problem was posed, therefore, involved changing its presupposition to: Given that chemical reactions are causally governed by affinities, temperature, and concentration, why does this chemical reaction yield these products instead of these other products?

The second type of improvement involves a better specification of what is being explained. We can think of the alternatives as defining a space of possible contrasts subdivided into areas within which different possibilities are only *insignificantly different*. This makes an explanation stable: small differences in what is to be explained do not invalidate it. On the other hand, possibilities must differ significantly from each other across areas, this being what makes them relevant alternatives.[44] In the case of chemical reactions, the differences that matter are differences in composition, since these determine differences in properties and hence in dispositions to combine, so the improvement was made possible by quantitative statements about composition. An example comes from the closing years of the 1700s, when it was discovered that several metals reacted with oxygen to form alternative compounds. For example, if 8 grams of oxygen were reacted with 31.8 grams of copper, the product was black copper oxide. But if the amount of copper was 63.6 grams, then the product was red copper oxide.[45] This outcome was clearly in need of explanation, but *not* an explanation given as an answer to the question "Why does the reaction of oxygen and copper yield copper oxide, instead of other compounds?" but rather: Why does this reaction of oxygen and copper yield black copper oxide, instead of red copper oxide?

Whereas the contrast space of the former question is divided into two areas, the area containing substances insignificantly different in composition from copper oxide and the area with all other substances, the second question subdivides the first area into two sub-areas, each one with products with the same composition but different proportions. Thus, the contrast space becomes more finely subdivided, and the solution to the problem better specified, by using both composition and proportions.

Explanatory schemas share statements with problems, so ways of making the latter better can also be expressed as improvements in the former. Thus, the improvement in presuppositions just mentioned could be described as a change in the Affinity schema, from statement A-4 to statement A-4':

A-4. The affinity of a substance for another substance is constant.
A-4'. The affinity of a substance for another substance can change depending on the effect of a variety of other factors, including temperature and concentration.

Similar improvements occurred to the Part-to-Whole schema. Two of its statements had been in constant tension throughout the 1700s:

PW-3. The most basic components are those that cannot be further decomposed using existing chemical operations.
PW-4. The most basic components are few in number (four or five principles).

For these two statements to be compatible, the basic principles had to be the limit of chemical analysis *in practice*, but some of them, like the Sulphur principle, had resisted isolation by analytical techniques, and others (Water, Air) had been shown to be compounds before the century was over.[46] Conversely, a growing number of newly isolated substances defied further analysis, so to be consistent with PW-3, had to be declared elementary. This tension eventually forced chemists to accept that the number of basic building blocks was an empirical question. This change can be expressed as the replacement of PW-4 by PW-4':

PW-4'. The number of basic components is determined empirically.

Another statement, PW-2, was changed as a response to a better understanding of the difference between mixtures of two substances and the compounds these substance formed when united. In a mixture there is continuity between the properties of the components and those of the whole, so that adding a larger quantity of a principle like Water simply made the whole more fluid, while adding to it more Earth, made it more solid. But when a new compound was formed under the action of affinity—when a neutral salt was produced from the union of an acid and an alkali, for instance—it was easy to verify that the whole had different properties than its parts. Thus, statement PW-2 had to be changed to PW-2':

PW-2. The properties of a whole are a blend of the properties of its components.
PW-2'. The properties of a whole are a blend of the properties of its components in the case of mixtures, or a set of properties different from those of its components in the case of compounds.[47]

If the previous argument is correct, and cognitive tools can indeed be improved according to their own criteria and in response to their own demands for change, it is conceivable that a series of such improvements can lead to the betterment of the personal practices of chemists. On the other hand, this argument does not show that such advancements can spread throughout the entire community and become *consolidated*, so that the cognitive gains can be prevented from dissipating as individual practitioners die or lose influence in their communities. This loss can be forestalled by the incorporation of elements from personal practices into *consensus practice*, codified into textbooks and passed from veterans to apprentices. The consensus practice at any one point in time consists of the same cognitive tools as before except that they have become more *impersonal*. Whereas in personal practices a belief in the truth of certain statement, or the significance of a certain problem, or the productivity of certain taxonomic schema, may not be separable from the reputation of the practitioner who first stated it, posed it, or devised it; once these components are shared they become separated from their source. The basic question of *social epistemology* can then be formulated as follows: Can changes in the personal practices of an entire community result in a sequence of consensus practices in which the different components can be shown to improve over time?[48]

Circumstances that would prevent personal gains spreading and becoming part of the practices of the community are not hard to imagine: an authoritarian gate-keeper, like the editor of a journal or the head of a learned society, blocking the entry of certain achievements; the outcome of a power struggle determining what is deemed not to have sufficient merit to become part of a new consensus; the absence of forums within which debate can be carried out using arguments and evidence. A full discussion of these and other social factors will be postponed until the final chapter, but we can get the argument started by pointing out how conceiving of the cognitive content of a field as a heterogeneous set of tools can make the formation of consensus more likely. When a debate is modeled as involving the clash of two monolithic theories, it is easy to conclude that a complete breakdown in communication will occur: if two sub-communities disagreed about the reference of every concept, the truth of every statement, the significance of every problem, or the value of every explanatory or taxonomic schema, then reaching agreement through argumentation would be impossible, and a switch from one rival theory to another would be like religious conversion.[49] But if, on the contrary, each cognitive tool changes at its own rate, and if the changes are distributed differently across a community, then the existence of *multiple partial overlaps* becomes plausible: some practitioners may disagree on whether a concept has a referent but agree on the truth of statements in which the concept appears; or they may dispute the value of an explanatory schema but accept the significance of the problem the schema is meant to solve. These overlaps can keep communication from breaking down during a controversy, so that mutual understanding of arguments in favor of and against the disputed conclusions are possible, and new evidence bearing on the dispute can be taken into account.

To illustrate this approach we need a way to model contentious episodes in a way that displays their *temporal structure*, and then show how this model can be applied to a well-known controversy from the eighteenth century. The first feature of the model is a graph, referred to as an *escape tree*, showing the alternative moves open to the participants at any one time, along with the cognitive costs and benefits involved in the acceptance or rejection of a given alternative.[50] The graph allows us to follow the changes in the alternatives as arguments become refined, new evidence produced, and awareness of costs and benefits increased. The second feature of this model is the assumption that in a community in which each practitioner has limited cognitive abilities, a restricted access to relevant information, and only a partial understanding of the consequences of each alternative, the space of possibilities *can only be explored collectively*. As participants develop new arguments or produce evidence that blocks one branch of the escape tree, the space of possible alternatives changes and credibility is redistributed. And despite the fact that none of the arguments used to block branches of the escape tree can compel agreement, *a collective line of argument can emerge* that most participants agree is highly compelling, even if the agreement is only retrospective.[51]

The third feature of this model is the *constructive role played by dissenters*. Practitioners who object to concepts, statements, problems, classifications, or explanations that, in retrospect, can be judged to be improvements over past practice, can still play a role as critics, revealing faults in the arguments of their rivals; raising awareness of the costs of accepting their alternatives; and defending pieces of contrary evidence. Because the emergent line of argument is slowly produced as the space of alternatives is collectively explored, it is important that the variability in personal practices is maintained and that consensus is not reached prematurely, even if what prevents agreement from forming are factors like shared nationality, personal animosities, professional ambition, or just plain stubbornness.[52] Once a controversy is settled, some of these non-cognitive factors may outlive their usefulness. Nevertheless, it may still be reasonable for participants on the losing side to hold on to their views because cognitive gains in some components are compatible with *cognitive losses* in others: the controversy may have involved several unsolved significant problems, for example, only some of which had been solved at the end, while the remaining ones may have been unfairly discarded.[53] This implies that cognitive gains need not accumulate in a linear and uniform way. All that matters is that gains outweigh losses, and that the legitimacy of the emergent argument outweighs the effects of authority and political alliances.

Let's apply this model to a famous controversy that took place in the late 1700s. This episode is usually characterized as a discontinuous change from one monolithic world-view to another: a world-view based on oxygen against the one based on the Sulphur principle, or as the latter was renamed early in the century, *phlogiston*. In this characterization, chemists who explained combustion phenomena using oxygen and those who used phlogiston lived literally in different worlds.[54] A more careful examination of the controversy, on the other hand, reveals that it had a complex structure: it lasted a quarter of a century; it involved such a large portion of the domain that none of the participants had expertise in all relevant phenomena; it changed in nature as new evidence was added; and the set of possible solutions was explored collectively, with critics playing a constructive role in the creation of the final argument; and the participants shared enough common ground that a complete breakdown in communication never happened. The concept of affinity, for example, was never questioned by either side, nor was the data about displacement reactions contained in the affinity tables, so the resolution of the debate left this part of the domain mostly untouched.[55]

Let's start by describing the relatively large portion of the domain that was under scrutiny. The explanation of several chemical reactions was contentious: *calcination*, the analytic transformation of a solid and shiny metal into a dusty and opaque calx; *reduction*, the transformation performing the inverse operation, converting a metallic calx into a solid metal, usually with charcoal operating as reducing material; and finally, *acidification*, the transformation undergone by metals in interaction with acids, an apparently unrelated type of chemical reaction until chemists realized that it could be conceived as a form of liquid calcination.[56] The number of phenomena associated with these three types of reaction was so large that mastering them all would have tested the abilities of even the best chemists at the time. To make things even more complicated for the participants, some of the substances involved were newly discovered "airs," the status of which was itself unsettled. We mentioned above the dilemma posed by incorporeal substances (air, fire, light, magnetism) and said that using the pneumatic trough to capture the "airs" produced during a chemical reaction had allowed their eventual incorporation into the domain. But as these previously neglected reaction products were studied, differences in their chemical behavior were noticed, leading to a profound conceptual change: first came pluralization, a move from "air" to "airs," as a variety of gaseous substances were isolated; then as their affinities with other substances were established, it was realized that they differed in chemical nature, that is, that they were different substances; finally, chemists had to perform a mental leap from the idea of air as a basic substance to that of a *state of matter* that could be displayed by many substances.[57] Only then was the term "gas" revived to refer to the elastic fluid state itself.[58]

Given the large portion of the domain involved, and the fact that the very chemical identity of the newest addition to the domain was in question, it is not surprising that for every phenomenon there were many rival explanations, none of which could compel assent.[59] This situation is precisely of the type that could benefit from a collective exploration of the escape tree. There were seven practitioners directly involved—four French, two British, and one Irish—with several others contributing from the sidelines.[60] Despite their disagreements, these seven chemists were involved in a collective exploration of the *same* space because of what they shared in common. The British, who ended up on the losing side, were the masters of *pneumatic chemistry*, the branch of the field that contributed the cognitive tools and instrumentation that would eventually decide the outcome. The French chemists had to catch up with them as the debate raged, benefiting from their successes and learning from their failures.[61] This means not only that all participants shared access to the resources needed to solve the problem that triggered the controversy, but that they were all trying to solve one and the same problem, the problem expressed by the following question: Given that calcination is a process in which one component of metallic substances (phlogiston) is *emitted and lost*, why are the calces produced by it heavier rather than lighter?

The phenomenon itself had been noticed since early in the century, but in only a few cases was it remarkable enough to focus attention. It was well known, for example, that lead gained a surprising amount of weight after being roasted. But it was not considered an urgent problem until 1771, when a detailed study determined unambiguously that the weight gain was indeed a general phenomenon affecting many metallic substances.[62] The formulation of the problem just given shows its main presupposition but it will be useful to state all of them more explicitly.

The set of statements presupposed by the weight-gain problem was the following:

1   The direction of any reaction, as well as the nature of the final products, is caused by the affinities of the different substances involved.

2   The overall amount of matter at the start of the reaction remains

the same at the end of the reaction. Hence the weight of the reactants must match the weight of the products.

3   The air in which the calcination reaction takes place plays only an auxiliary physical role, not a chemical role.

4   Calcination is an analytical operation that separates a metallic substance into its two components:

   a   The calx that remains in the vessel.

   b   The phlogiston that is emitted.

5   The emitted phlogiston:

   a   causes the fire and light observed during calcination.

   b   causes the loss of metallicity.

All participants accepted statements 1 and 2, ensuring continuity as the weight-gain problem was debated.[63] The other statements underwent a series of challenges, but these remained part of the personal practice of a single participant for a full decade.[64] In 1773, for example, calcination experiments in a closed vessel showed that the weight gain by the calx was the same as a the weight loss in the air around it. If this could be confirmed, it strongly suggested that air becomes solidified or "fixed" into the calx as the combustion proceeds, accounting for the increase in weight. This insight reinforced a previous result involving the opposite transformation: reducing lead calx to restore its metallicity liberated a large volume of air at the moment that the transition from calx to metal occurred.[65] These two results presented a challenge to statement 4b, and more generally, to a conception of calcination as a reaction in which something is emitted. Another contemporary series of experiments on the reversible transition of vapor to liquid water not only strengthened the insight that air could change state and become less volatile, but also that the difference between a liquid and its vapor was the possession of the matter of fire in greater quantity. This could be used to challenge statement 5a, since the fire and light observed during calcination could now be conceived as being caused by a component of the air not of the metal.[66]

At this point in the controversy the available evidence could not determine the choice of explanations. It was entirely reasonable for all participants to think that there were alternative branches of the escape tree that were compatible with the evidence. Or if the existing branches seemed inadequate, a participant could add a new one. On the basis of the observed volatility of fumes and other products of combustion, for example, a new presupposition could be added: that *phlogiston is lighter than air*, an assumption which, if true, would imply that when the matter of fire is fixed in a metal it makes it more buoyant relative to the air, much as objects that are lighter than water (corks) can make an object that is heavier (a fishing net) float.[67] Early on in the debate, adding this presupposition was relatively costless, but as it progressed the cognitive costs increased. As the different products of combustion reactions were carefully quantified, and the weight of the gases estimated, the added assumption raised a question: why should the weight of the air in a closed vessel diminish if it has lost matter of fire to the metal? And more importantly, this added branch to the escape tree could easily be blocked by weighing reactants and products in a vacuum.[68]

By 1780, most of the participants had accepted that something was indeed absorbed, and that this is what explained the problematic increase in weight. But statement 4b could still be saved if another presupposition was added to statement 5, stating that the emitted phlogiston:

5c combines with the surrounding air to form a new substance, a substance that is absorbed by the calx.

This alternative branch of the escape tree had opened thanks to two developments. One was the proliferation of new "airs": fixed air (carbon dioxide); inflammable air (hydrogen); and vital air (oxygen). News about these potentially new substances spread from Britain to France, where chemists rushed to master the laboratory procedures to isolate them, and to repeat the experiments that had been performed on them. The second development was a remarkable phenomenon first reported in 1784: using an electric spark, a mixture of hydrogen and oxygen in the right proportions could be detonated to produce water. These two results strongly suggested that *phlogiston was in fact hydrogen*: that in fixed or solidified form it was a component of metals; that when released it combined with the oxygen in the air to form water; and finally, that this water became absorbed in the calx, explaining its gain in weight.[69] But adding 5c carried significant cognitive costs for those trying to save 4b. In particular, those defending the existence of phlogiston as a basic principle were also committed to defend other basic principles, such as water, but using 5c as part of an explanation of the weight-gain phenomenon implied accepting that water was a *compound*. Those who found this unacceptable attempted another way out: water's status as a principle could be maintained by arguing that hydrogen was water saturated with phlogiston, while oxygen was water deprived of phlogiston. The problematic experiment, therefore, was not a synthesis of water at all; it was more like mixing two impure forms of elementary water.[70]

Several more moves and countermoves were attempted, but despite the fact that the escape tree still had several branches open, these were all now *analytically contingent*. That is, given the shared assumption that an elementary substance is whatever cannot be decomposed by purely chemical means, the status of air and water as undecomposable was contingent on the improvement of analytical techniques, such as the advent of electrolysis, which a few decades later would make breaking down water fairly routine. Indeed, towards the end of the controversy, the main source of disagreement had been boiled down to this: what was considered an elementary substance by one side was a compound for the other.[71] Unlike the previous sources of contention, which ranged over a variety of issues, this one could be decided experimentally, making the state of underdetermination transitory.[72] One by one, the defenders of phlogiston became convinced that the costs of believing the statements in which this hypothetical substance figured outweighed the benefits, and joined the other side. These defections took place not as a religious conversion from one world-view to another, but as a result of *protracted argumentation*.[73]

The question of whether improvements in personal practices can enrich consensus practice is contingent: it is certainly possible that the historical record contains controversies in which there are more losses than gains. Hence, it is important to perform the analysis of costs and benefits correctly. It has become routine, for example, to assert that those who supported phlogiston, but not their rivals, had a correct solution to the following problem: Why do all metallic substances have similar properties?[74]

If this were a valid thesis then the resolution of the controversy would have involved the loss of a correctly posed problem and of its true solution: because all metals are compounds of a calx and phlogiston. But some chemists earlier in the century had already questioned the truth of this statement because they had not been able to decompose metals any further. And the next generation of chemists, using improved analytical techniques, also failed to decompose them. So the statement "All metals are elementary substances" received further confirmation as time went on. It was only by the middle of the nineteenth century, when many new elementary substances had been isolated, identified, and arranged in a table that displayed their regularities—including the regular arrangement of metallic substances relative to other substances—that the problem of why they all share certain properties could be correctly posed. This does not imply, on the other hand, that all the statements in which the concept of phlogiston had been replaced by that of oxygen were correct. In fact, some key statements held to be true by the winners of the controversy were strictly speaking false:

6 Oxygen is a component of all acid substances.

7 Oxygen is necessary for all combustion reactions.

8 Oxygen gas is a compound of solid oxygen and caloric (the matter of heat).

In the 1790s, acids not containing oxygen were isolated, falsifying statement 6, although no one paid much attention at the time.[75] In the first decade of the next century, combustion—a chemical reaction in which there is emission of heat and flame—was shown to occur in other gases, like chlorine, limiting the truth of statement 7. And the idea that gases were compounds, or equivalently, that heat is a substance, would soon be abandoned, although in this case the advance was not made by chemists.[76] Given that these three statements were shown to be false (or only partially true) when the controversy had not yet been fully decided, it would have been perfectly reasonable for a chemist at the time to stick to phlogiston. On the other hand, all three statements had some truth to them—many acids do contain oxygen; many forms of combustion do require oxygen; the gas phase does require the presence of heat—which means that the statements were *improvable*. This information was not available to chemists in 1800, but it was to those working in 1850. And those future chemists had access to another piece of information: oxygen had resisted all attempts at decomposition made in the intervening 50 years. Given the analytical contingency of the definition of elementary substance, the fact that the status of oxygen has not changed even today should be part of the track record of this once hypothetical substance, a track record that should certainly be taken into account when giving an answer to the following question: Is an ontology based on oxygen an improvement over one based on phlogiston?

The answer to this question is not straightforward because the term "phlogiston" had several referents. The original referent, the old Sulphur principle, turned out not to exist by chemistry's own criteria: it never was a product obtained as the limit of a chemical analysis. By 1750, chemists had proposed that its referent was the matter of heat, fire, or light. If this were correct then it can be argued that phlogiston survived until the end of the controversy, but that it had been relabeled "caloric." This is only partly true: the term "caloric" referred to the substance emitted during combustion by *all* burning substances, not only those believed to possess the inflammable principle.[77] Finally, if we take the referent of the term "phlogiston" to be inflammable air, that is, hydrogen, then whether there was an ontological improvement as a result of the controversy can be settled by using chemical analysis to answer questions like: Is water an elementary substance? Are pure metals compound substances? By 1800, most textbooks were answering these questions in the negative, and this consensus has lasted till the present day. Given this track record, we can state unambiguously that replacing phlogiston by oxygen constituted a great improvement.

From Personal to Consensus Practice 1700–1800

1700–50

Georg Ernst Stahl (1659–1734), Wilhem Homberg (1652–1715), Etienne-François Geoffroy (1672–1731), Hermann Boerhaave (1668–1738), Stephen Hales (1677–1761), Gillaume Francois Rouelle (1703–70), Pierre Joseph Macquer (1718–84)

The domain of chemistry in 1700 did not have sharp boundaries: it blended gradually into the realms of craft and trade. Many substances found in laboratories were neither from natural sources nor created by chemists. Rather, they were traditional substances produced in mines, apothecary shops, or distilleries, and commonly traded as commodities: gold, silver, copper, tin, lead (from metallurgy); mercury, antinomy, and arsenic (from pharmacy and medicine); potash and soda (soap and glass manufacturing); nitric, sulphuric, and muriatic acids (from a variety of crafts). Even the raw materials for vegetable extracts came from medical and botanical gardens, not from Nature.[1] The other component of the domain, chemical reactions, had also been part of the material culture of assayers, apothecaries, and distillers for centuries. Even the tools and instruments used by chemists were not distinguishable from those used in the crafts, and they were typically obtained from the same providers.[2]

Given this relation with the material culture of better established practices, it is not surprising that part of what had become consensus in early chemistry came from what had been traditionally agreed upon in those other areas. The concepts designated by the terms "metal," "acid," "alkali," "salt," were well established. Statements that contained these concepts, statements like "Two metals combine to form an alloy" or "An acid and an alkali react violently and are transformed into a neutral salt," had been considered true for a long time.[3] Classifications of substances into Animal, Vegetable, and Mineral categories were customary and in widespread use. But early chemistry textbooks, produced in countries with a strong didactic tradition, showed that the consensus practice of chemists had already acquired a few distinguishing features. The most important chemical reactions in this century, displacement reactions, although already known to metallurgists, were featured in textbooks as *recognized empirical problems*: the selectivity of the reactions that the replacement of one dissolved substance by another revealed had become legitimate targets for explanation, something they were not in the hands of artisans.[4]

Similarly, concepts from ancient philosophers and alchemists, like the concept of a basic element or principle, had been transformed as *the compositional problems* in which they figured changed. Air, Water, Fire, and Earth, together with Salt, Mercury, and Sulphur, were retained as the most basic components, but while before chemistry a compound made of these elements was viewed as a seamless totality in which the parts lost their identity, the ability of chemists to analyze a substance and use the products of analysis to resynthesize it and recover its original properties promoted the idea that its components retained their identity.[5] This new conception of the relations between parts and wholes had seventeenth-century precedents, but it only became widely accepted at the dawn of the eighteenth, when the analysis–synthesis cycle had become the very definition of chemistry.[6] The old principles had also changed in another way: *their referents were now fixed instrumentally*, at least as an ideal. Through the careful control of fire, the distillation of plant raw materials gave as products substances that were close in their sensible properties to those of the postulated principles and elements. Although these were not considered to be the principles themselves, they were treated as their legitimate instrumental counterparts.[7]

In the seventeenth century the art of distillation, and the use of fire as a solvent, had acquired the status of exemplary achievements, the very standard of what an analytical technique could achieve. To sustain this line of research, furnaces of increasing power were built.[8] Nevertheless, because at this stage in its evolution chemistry needed to legitimize itself by its medical and pharmaceutical applications, dry distillation could be challenged if it interfered with the preparation of remedies. In the early 1600s such a challenge had indeed been issued, and it gathered strength as the century came to a close. Some practitioners began to doubt, for example, whether fire recovered the original components of vegetable raw materials or whether its action was so strong that it produced new ones. Evidence began to accumulate that some substances created using fire as a solvent had lost their medicinal capacity to affect the human body. In response, some chemists proposed switching to wet distillation and the use of water as a solvent. By the early 1700s, solution analysis had joined dry distillation as a laboratory technique, not with the same degree of consensual approval but with an increasing appeal to those engaged in the preparation of pharmaceutical remedies.[9]

Solution analysis also had links to metallurgy, the craft in which displacement reactions had been first noticed.[10] As we argued in the previous section of this chapter, these reactions were regarded as a truly remarkable phenomenon, vividly displaying the progressively stronger affinities that metals had for particular acids in solution. In the case of nitric acid, for example, a well-defined sequence of affinities could be identified: silver, copper, iron, zinc, each one first dissolving in the acid then precipitating as the next metal was added to the mix. The use of synthesis to corroborate the results of analysis was also of metallurgical origin: if one could decompose an alloy like bronze into its constituent copper and tin, two substances with quite different properties, and then resynthesize bronze from these components, this proved not only that bronze was made out of copper and tin, but that the latter two metallic substances subsisted without loss of identity in the alloy. By contrast, while distillation could extract as many different substances from plants as there were basic principles—mineral raw materials yielded at most two principles—chemists could not use these extracts to resynthesize the original plant materials. Only reversible chemical transformations provided a secure ground for compositional problems, and they led the way as the new field searched for its identity.[11]

Thus, in 1700 a variety of cognitive tools shaped the collective practices of chemists, or more exactly, of the physicians, apothecaries, metallurgists, and natural philosophers who practiced chemistry at the time: general concepts for a variety of animal, vegetable, and mineral substances, as well as for their sensible (taste, smell, consistency) and measurable (weight, solubility) properties and dispositions; statements about the transformations that these substances undergo as they interact with one another; problems expressed by questions like "Why does this substance have a tendency to react with this substance instead of other substances?"; and finally, reasoning patterns that we can reconstruct using explanatory schemas, like the Part-to-Whole schema:

Question
Why does substance X have properties Y and Z instead of other properties?

Answer

PW-1. Properties are explained by the Part-to-Whole relationship between a substance and its components.

PW-2. The properties of a whole are a blend of the properties of its components.

PW-3. The most basic components are those that cannot be further decomposed using existing chemical operations.

PW-4. The most basic components are few in number (four, five, or six elementary principles).

PW-5. The properties of a whole are determined by the chemical nature of its components.

There was a contemporaneous variant of this schema, adopted from natural philosophers, in which the second statement was:

PW-2'. The properties of a whole are a product of the size, shape, motion, and arrangement of minute corpuscles.

Using this alternative version of the Part-to-Whole schema, the properties and dispositions of an acid substance—it pricked the tongue when tasted; it formed crystals with sharp edges; it had the capacity to dissolve metals—could be accounted for by postulating the existence of *pointy corpuscles*. These allowed an acid to act as solvent, as its sharp particles penetrated the pores of other substances, causing them to break apart, while their geometry and arrangement could explain why liquid acids solidified forming crystalline structures.[12]

Thus, within the consensus of 1700 there was room for variation at the level of personal practices. One source of variation was the influence of different traditional crafts: German chemists (Stahl, Homberg) were more influenced by mining and metallurgy than French ones (Geoffroy, Rouelle), who had more of a pharmaceutical background. A second source of variation was the attitude towards natural philosophy: some chemists (Stahl, Geoffroy, Rouelle) fearing reductionism drew a sharp line between chemistry and physics, while others (Homberg, Boerhaave, Hales) did not feel a threat and imported concepts and instruments from that field. A third source was the use of new instrumentation that either promised higher analytical resolution (Homberg, Geoffroy) or that transformed previous byproducts of chemical reactions into new phenomena (Hales). Finally, there were the various ways in which different practitioners dealt with the tension between an instrumental definition of principles as the limit of chemical analysis, and the hypothetical ancient principles that could not be isolated in the laboratory.

We begin our examination of personal practices with the work of Georg Stahl. This German physician was the center of a community of chemical analysts in Berlin in 1715, his ideas crossing national borders into France soon after that. His solution to the dilemma posed by the instrumental and hypothetical conceptions of the basic principles was perhaps his most influential contribution. Stahl thought of the substances in the chemical domain as forming a *compositional hierarchy*: at the lowest level there were the basic principles, inaccessible to the chemist in pure form; these principles combined with each other to form primary mixts, the simplest kind of substance that a chemist had direct access to; primary mixts, in turn, could be combined to produce secondary mixts, that is, compounds and other aggregates. This hierarchy allowed Stahl to maintain a belief in the existence of a few basic principles while denying that they had to be the limit of chemical analysis, a status now reserved for primary mixts.[13]

The basic principles that chemists accepted varied in number and kind. There were seven possibilities, four from Aristotle (Water, Air, Fire, Earth) and three from Paracelsus (Sulphur, Mercury, Salt). Stahl, following his teacher Johann Becher, believed in four principles: Water and three different Earths, sulphureous, mercurial, and vitrifiable. The three Earths carried the basic properties of inflammability, volatility, and solidity. He did not consider the two traditional elements left out, Fire and Air, capable of causing chemical change, only to enable it: Fire as an instrument, Air as an environment.[14] But Stahl broke with Becher in one important respect. Traditionally, the component responsible for metallic properties had been considered to be Mercury. But his experiments had revealed to him that *the corrosion of metals and the combustion of wood* were closely related phenomena. This implied that the loss of metallicity (by corrosion or calcination) was linked to the inflammability principle, that is, to Sulphur. Stahl changed the principles to conform to this belief and marked the change by giving a new name to the Sulphur principle: phlogiston. He was aware that the calcination of metals was a reversible reaction, that is, that the original metal could be synthesized by heating it in contact with charcoal, and this suggested a plausible mechanism: as phlogiston leaves a sample of shiny, solid metal it causes it to turn into a dull, though sometimes colored, dust; and vice versa, as it enters a metal calx from the charcoal, it restores its metallicity.[15]

The personal practice of Stahl's contemporary, Wilhem Homberg, was different in a number of ways. Homberg was more influenced by natural philosophy and was therefore more inclined to use instruments like the air-pump or the aerometer to measure physical properties like *specific weight*, the weight of a substance per unit of volume. Homberg pioneered the quantitative approach to chemistry when he set out to measure the strength of different acids by slowly adding an acid solution to a weighted quantity of alkali until effervescence stopped, then drying up and weighing the resulting product. He believed that the difference in weight of this final product compared to that of the alkali indicated the weight of the acid contained in the solution. Because at this point there was no way to capture and weigh the gases (carbon dioxide) produced by the reaction, his results were not correct.[16] But his procedure implied the belief that *weight is conserved in a chemical reaction*, a belief that would play a crucial historical role. Homberg was also the first to define chemical practice as the art of both analysis and synthesis, and to view synthesis as a means to confirm the results of analysis.[17]

Homberg's accepted five basic principles: Water, Earth, Sulphur, Mercury, and Salt. Like Stahl, he thought that these principles could not be isolated, but he dealt with the tension between the instrumental and the hypothetical by giving himself more freedom to speculate. This way he was able to move from the realm of actually achieved analysis to that of what analysis could achieve if its resolution were increased, and from there to the realm of invisible corpuscles.[18] Homberg pioneered the use of the *burning glass*, a convex glass three feet in diameter that was used to focus solar rays on a sample, achieving much higher temperatures than existing furnaces. This made focused sunlight a better solvent than fire, greatly increasing the resolution of distillation analysis. He used the glass to burn different metallic substances in an attempt to isolate their most important component, the Sulphur principle. In the process, he came to believe that the light he was manipulating played a chemical role, entering into composition with metals and accounting for the gain in the weight of their calces. In other words, he identified the Sulphur principle with the matter of light, which combined with inflammable oil became a flame. Decades later, French chemists would adopt a similar position when they redefined phlogiston as the matter of fire.[19]

The personal practice of Homberg's student, the French chemist Etienne-François Geoffroy, combined elements from the previous two variants. Like his teacher, Geoffroy used the burning glass to perform analyses and resyntheses of iron, copper, lead, and tin. But because he was an apothecary he was much less inclined than a natural philosopher to engage in speculation about invisible corpuscles. He accepted the same five principles as Homberg, but adopted Stahl's conception of the Sulphur principle: it was the principle responsible for metallicity, and what was emitted by metals during calcination. Unlike his predecessors, however, he did not think that Sulphur could be studied only indirectly, by the activity it conferred to other substances. In fact, he believed he had actually succeeded in isolating it when he produced a thick oily substance during his burning glass experiments.[20] Geoffroy broke with his teacher in another respect, focusing research on chemical rather than physical properties. Thus, while Homberg had approached the problem of displacement reactions by quantifying the strengths of acids, a line of research that could not address the question of *selectivity*, Geoffroy saw the latter as a uniquely chemical property. But believing that speculation about the mechanisms of selectivity was premature, he set out instead to organize the growing body of statements about displacement reactions in tabular form. Geoffroy's table of affinities of 1718 went beyond a mere collection of empirical observations in two different ways. First of all, the table brought order to the chemical domain by gathering together all the known ingredients of neutral salts, not just acids and alkalis but also metals.[21] And second, the table supplied chemists with another way of dealing with the tension between the hypothetical and the instrumental: it became an *unofficial list of elementary substances*, its contents being more decisive in defining the proper level of ultimate analysis than any set of ancient principles.[22]

The personal practice of Hermann Boerhaave, Geoffroy's Dutch contemporary, was shaped to a greater extent by natural philosophy. His laboratory, like that of Homberg's, contained instruments borrowed from that other field, most notably the thermometer. Thermometers were not stable instruments in the 1700s. Their standards were in flux—what expansible liquid to use, mercury or alcohol? What graduated scale to use?—as were the methods of calibration needed to make measurements performed in different laboratories comparable to each other. Even the conceptual basis of thermometry, specifically, the distinction between the concepts of *amount of heat* and *intensity of heat* (temperature), was far from settled. These difficulties were not solved in Boerhaave's time, but his practice contributed to the development of the instrument itself, through his association with Daniel Fahrenheit, and to the ontology of heat and fire.[23] The basis of the thermometer, the phenomenon of the expandability and contractibility of a substance under the influence of heat, led Boerhaave to think about composition in terms of *forces of attraction* (leading to contraction) and *repulsion* (leading to expansion). And this, in turn, allowed him to add one more variant to the Affinity schema we discussed in the previous section of this chapter. Let's recall the schema:

Question

Why does the reaction of substance X with substance Y have substance Z as its product, instead of other products?

Answer

A-1. Substances possess dispositions to selectively combine with other substances.

A-2. If X and Y are elementary substances then their dispositions will determine Z: if they have affinity for each other their product will be a combination, else they will not combine.

A-3. If X or Y are compound substances then the dispositions of their components will determine Z:

> A-3a. If only one of the substances is a compound, then the elementary substance's affinity for one of the compound's components will cause it to be dislodged and join with it, displacing the other component.

> A-3b. If both substances are compounds, then they will both decompose and their components will unite following their respective affinities.

A-4. The affinity of a substance for another substance is constant.

A-5. Two substances have affinity for each other if they display compositional similarity: two substances composed of the Sulphur principle, for example, will tend to combine with each other.

Most chemists of this period accepted the first four statements of the schema, but depending on the version of Part-to-Whole to which they subscribed, they would accept different versions of statement A-5. Stahl, who accepted the schema Part-to-Whole with statement PW-2, subscribed to the Affinity schema with statement A-5. Homberg flirted with a version containing statement PW-2', so if pressed he may have accepted something like:

> A-5'. Two substances have affinity for each other if their particles display a correspondence between their shapes or motions: acids have a great affinity for alkalis because the pointy shapes of their particles fit the pores formed in globules of alkali's particles.

Boerhaave's studies of the effects of heat on substances led him to introduce a Newtonian variant to rival the Cartesian statement A-5':

> A-5". Two substances have affinity for each other if there exists a force of attraction between them, and if the force is strong enough to overcome any existing forces counteracting attraction.[24]

As Boerhaave was taking the initial steps in the long process of elucidating the ontology of Fire, the English chemist Stephen Hales was pioneering the study of another incorporeal: Air. Hales began a line of research that others would later bring to fruition in what eventually came to be known as *pneumatic chemistry*. He invented an apparatus allowing the "airs" produced in the course of a chemical reaction to be captured, the initial step in making gases part of the domain. On the other hand, he was only interested in the quantitative study of the physical properties of the gases he collected, and this led him to overlook their differences in chemical properties. Retrospectively, we can infer from his experiments that he actually isolated hydrogen and oxygen, but he threw away these gases after having determined their quantity, failing to realize that they were distinct substances.[25] Nevertheless, he made the crucial discovery that would eventually lead to the recognition of the chemical role played by air: the plant materials he studied contained large amounts of air in a *fixed state*, air that could be released from them. This mattered because in this fixed state air could be thought of as a component of vegetable substances, not just as a neutral matrix enabling chemical change.[26]

The practitioners reviewed so far formed the first two generations of eighteenth-century chemistry. The following generation confronted a much richer domain, and a variety of novel cognitive tools, and thus felt confident that a unified account of the previous achievements could be performed. The work of Gillaume Francois Rouelle in the 1740s was an attempt at such a synthesis. Working in the French tradition, Rouelle continued the study of neutral salts pioneered by Homberg, but he could now attempt to perform an exhaustive classification. This was a different but complementary project to that of Geoffroy's, because while the table of affinities classified the components of neutral salts, Rouelle classified the products of the union of these components. He used properties and dispositions linked to the process of *crystallization* as a guide for his taxonomy: he first subdivided neutral salts into two classes on the basis of the amount of water needed to dissolve their crystals, solubility being correlated with the amount of acid they contained; then, he subdivided these classes by measuring the amount of water that entered into crystal formation; finally, he classified the ambiguous cases using the shape of the salt crystals. Rouelle adopted the Aristotelian distinction between genera and species to formalize his classification: the acid component was the genus, the base the species.[27]

Rouelle also reached outside the French tradition to bring into his synthesis concepts from Stahl, Boerhaave, and Hales. Fire, as well as the other principles, were treated by him as both components of substances and instruments of chemical change. Fire as a component was modeled on Stahl's phlogiston, so that the latter became the *matter of fire* when it was adopted by French chemists.[28] To conceptualize the role of fire as an instrument, Rouelle used Boerhaave's idea of heat as a force, the presence of which caused the expansion of substances, while its absence caused their contraction. Air was also given this double status: the apparatus of Hales was familiar to Rouelle and this allowed him to introduce not only pneumatic chemistry into France, but also to conceive of air for the first time as a substance that could perform a chemical role as a component of other substances.[29] Rouelle was a great teacher but not a good writer, so the synthesis of ideas he achieved affected consensus practice only when one of his students, Pierre Joseph Macquer, gave it the form of a general textbook.

Macquer accepted the four Aristotelian elements (Fire, Water, Air, and Earth) but wanted their definition to be instrumental, as the limit of chemical analysis, and analytically contingent: the four elements were to be considered basic building blocks only until someone succeeded at decomposing them.[30] He adopted Rouelle's conception of phlogiston as the matter of fire, but distinguished between free fire, as a component of any fluid substance, and fixed fire, or phlogiston proper, as a component of inflammable substances, acids, and metals.[31] He also included in his synthesis a subject that Rouelle had neglected: affinity. Like Boerhaave, he conceived of affinity as a force, but he also preserved something from Stahl's conception based on similarity of composition. To bridge the gap between the two he created a typology of different kinds of affinity: the affinity that a substance has for another sample of the *same* substance was distinguished from the affinity between two different substances.

These two categories were referred to as "affinity of aggregation" and "affinity of composition" respectively. A third category, named "affinity of decomposition," was used to classify displacement reactions, that is, the case in which a substance had a strong affinity for only one of the components of a compound substance, forcing it to break away and then uniting with it. His typology included other categories to cover the more complex cases of affinity that had been added to the domain since the publication of the original table.[32]

Macquer published a textbook in 1749 that was read and respected throughout Europe.[33] From its contents we can get a good idea of what concepts, statements, problems, taxonomic and explanatory schemas had been added to the consensus of 1700. His textbook is a synthesis of contributions from all the chemists just discussed. From Stahl, Macquer took the concept of a *compositional hierarchy*: chemists differed in what they considered fundamental principles, but they all agreed on the list of *secondary principles* that formed the next level in the hierarchy—acidic, alkaline, and neutral salts; a variety of mineral, vegetable, and animal oils;

and a diversity of metals. Compounds are assumed to be made out of these secondary principles. Thus, the original role of the hierarchy—to draw a wedge between the hypothetical and the instrumental—has become part of consensus practice.[34] From Homberg's experiments the book adopts the concept of *saturation point*: the proportions of acid and alkali that yield a substance that does not possess any acidic or alkaline properties, that is, a neutral salt.[35] This is an improvement because the referent of the term "saline substance" was not well defined, the term referring to any substance that possessed properties that were a blend of those of Water and Earth, including acidic and alkaline substances.[36] The concept of saturation point, and the quantitative approach that it made possible, allowed fixing the referent of the term "neutral salt" with more precision. In addition, as Rouelle had shown, the different neutral salts could be distinguished from each other through a determination of their solubility, their ability to form crystals, the shape of their crystals, and the ease with which the crystals melted when exposed to fire.[37] This too represented an improvement in the way in which the referent of the general terms for different neutral salts was fixed.

Chemists inherited from metallurgists reversible transformations, like those of calcination and reduction, conceived as ways of assembling and disassembling metallic substances.[38] By 1750 this reversibility has been shown to exist in many non-precious metals, and substances needed for the recovery of metallicity (reducing fluxes) have been identified.[39] Reversible operations were also known in pharmacy, a craft with which the study of neutral salts kept a close connection. Yet, although chemists were interested in medicinal applications, neutral salts acquired a more important role as exemplary cases of non-metallic substances that could be analyzed and synthesized routinely.[40] The explanations given in the textbook for these reversible chemical reactions are all framed in terms of the concept of *affinity*, evidence that the concept has become part of the consensus, even though its name was still in flux and its causes not well understood.[41] Given the central role played by neutral salts, it is not surprising that among the statements that had entered consensus practice a large number were reports of analytical results about the composition of these substances. Examples of these statements are:

1 Glauber's salt is composed of vitriol and the base of sea salt.

2 Saltpeter is composed of nitrous acid and a fixed alkali.

3 Alum is composed of vitriolic acid and a vitrifiable earth.

4 Selenite is composed of vitriolic acid and the calx of stone.[42]

From statements like 1 to 4, Rouelle had inductively derived a very important generalization, expressed by statement 5. We may conjecture that this statement was ripe to be added to the consensus of 1750, but neither it nor the concept of a *base* (any fixed alkali, absorbent earth, or metal calx) appear in the textbook:

5 All neutral salts are composed of an acid and a base.

The most important taxonomic schema to enter the consensus in this 50-year period was undoubtedly the table of affinities. Twenty-nine different tables were produced after 1718, but only a single one had been published by the time the original version appeared in this textbook. But its inclusion stimulated a short burst of activity between 1769 and 1775 that produced seven improved variants.[43] That the table was improvable, however, is recognized in this textbook: newly discovered affinities have not yet been tabulated, and exceptions to the old ones are now known, but their the entries in the table have not yet been fixed. Macquer does not attempt to update the original table because, as he acknowledges, the recently produced information has not been sufficiently verified and the details are bound to be contested.[44] An important new problem, one that Stahl had first posed, was the transformation undergone by metals under combustion. In the decades after its first formulation, the problem had become recognized by most members of the community as being highly significant: Why are calcined metals dusty and matte, rather than ductile and shiny?

The consensus answer in 1750 was: because calcined metals have lost the component that gives them their metallicity. This problem had two presuppositions: that metallic substances were a compound of a calx and phlogiston; and that the application of fire to a piece of metal forced the phlogiston out (due to its greater affinity for the flame), leaving the powdery calx behind. The problem, its presuppositions, and its solution are enshrined by the textbook.[45] The consensus solution, on the other hand, raised other problems. A calx was considered to be an elementary substance, an Earth with the capacity to turn into glass. But if this was so then other vitrifiable earths, such as common sand, should turn metallic when phlogiston from charcoal was added to them. Yet, this had never been observed. Hence the problem was: Why don't all vitrifiable earths become metallic when compounded with phlogiston?[46]

Macquer's work is considered to be the inflection point at which the influence of Newton on French chemistry began to overshadow that of Descartes.[47] This change affected the fate of the different variants of the two explanatory schemas we have used to reconstruct the reasoning patterns of eighteenth-century chemists: Part-to-Whole and Affinity. In this textbook there is almost no speculation about the shape and motion of invisible particles, and explanations in terms of forces are beginning to be embraced. This shows that statement PW-2' has lost credibility:

PW-2'. The properties of a whole are a product of the size, shape, motion, and arrangement of minute corpuscles.

Statement PW-2, asserting that the properties of a substance are a blend of those of its components is retained only for ultimate principles, because given the latter's resistance to isolation and purification, the only way chemists could have knowledge about them was by observing their properties indirectly, in the blends they produced. Compound substances higher in the compositional hierarchy, on the other hand, contradicted this statement: the properties of a neutral salt are not a blend of acidic and alkaline properties, but *a new set of properties*. So while statement PW-2 is still part of the consensus, it is poised for future destabilization. The variants of the Affinity schema have also been affected: Cartesian versions using statement A-5' have been eliminated from consideration, but the Newtonian version with statement A-5" has not yet been fully accepted:

A-5'. Two substances have affinity for each other if their particles display a correspondence between their shapes or motions.
A-5". Two substances have affinity for each other if there exists a force of attraction between them, and if the force is strong enough to overcome any existing forces counteracting attraction.

So at this point in time, explanations using affinity still tend to assume that:

A-5. Two substances have affinity for each other if they display compositional similarity.

From the point of view of our own century, the most striking thing about the 1750 consensus is how recognizable it is as chemical practice: none of the concepts, statements, problems, taxonomic and explanatory schemas seems entirely alien to us. The nomenclature is strange, but the instrumental definition of substances makes it simple to demonstrate that a statement like "Glauber's salt is composed of vitriol and the base of sea salt" does not mark an unbridgeable discontinuity, since it can be easily translated into: Sodium sulfate is composed of sulfuric acid and sodium hydroxide.

The main difference between the chemistry then and now seems to be that what at the time was conceived as a compound (iron, copper) we treat as an elementary substance, and what they conceived as elementary (iron oxide, copper oxide) we treat as compound. Yet, this disagreement is easy to spot and does not create any incommensurability between their chemistry and ours.[48] Their ontology of Air and Fire may seem strange to us, until we realize how *objectively puzzling* incorporeal entities were at the time, and how many problems they would continue to pose for a century or more. Other parts of their consensus practice appear primitive, but they all seem eminently improvable. Finally, it is important to stress that prior to 1750 a *phlogistic theory* properly did not exist. Even though phlogiston was valued as the most active principle, it never formed the core of a theoretical system.[49] In fact, the only place we can find phlogiston playing a systematic role is in Macquer's textbook, but this means that the "theory" that later chemists rebelled against was a very recent one, one with which they shared many concepts, statements, and problems, and not the ancient one that can be traced to Paracelsus.

1750–1800

William Cullen (1710–90), Joseph Black (1728–99), Torbern Bergman (1735–84), Louis–Bernard Guyton de Morveau (1737–1816), Henry Cavendish (1731–1810), Joseph Priestley (1733–1804), Antoine-Laurent de Lavoisier (1743–94), Richard Kirwan (1733–1812), Claude-Louis Berthollet (1748–1822), Antoine Fourcroy (1755–1809)

Boerhaave taught chemistry to many Scottish practitioners, one of whom was William Cullen, Joseph Black's teacher. This lineage partly explains why Black's personal practice was focused on the ontology of heat and fire.[50] While Boerhaave focused on the rate at which heat was transferred from one substance to another, Black preferred to study the equilibrium state that was achieved once the heat transfer had stopped, much like Homberg had studied neutral salts at the point of saturation, that is, once the ebullition of the acid and alkali reaction had stopped.

Boerhaave thought that substances exchanged heat in proportion to their mass, that is, in a way determined only by their physical properties, but Black showed that chemical properties were at play, and hence that the kind of substance involved (not just its mass) was a significant variable. In this way he arrived at the concept of *specific heat*: the capacity that different substances have to absorb or release heat. Since at this point in time, heat is thought of as a substance, this capacity was explained using affinity: much like solid and liquid substances have different dispositions to enter into chemical combinations, so they have different affinities for heat. This implied that heat was a chemical entity, not just an instrument, as it was for Boerhaave.[51]

Black was helped in this research by the fact that the thermometer was now a stable instrument, with a standardized measuring scale and repeatable calibration procedures.[52] But his other main achievement was to break the link between the quantity measured by the thermometer and the amount of heat contained in a sample of a given substance. Cullen had already studied the phenomenon that gave rise to this distinction: the spontaneous cooling produced when a liquid evaporates. He studied the evaporation of 13 different liquids, alkaline and acid, and his results established the stability of the phenomenon.[53] Black went beyond this by observing that if heat was added to melting ice the result was not, as would be expected, an increase in the temperature of the ice, but only the production of more water. From this he concluded that the added heat became concealed, not only from the senses but from the thermometer itself, and that therefore the instrument did not measure the full amount of heat, only its intensity. This way Black arrived at the concept of *latent heat*, a concept that would have lasting consequences for the ontology of heat.

Black also contributed to deciphering the ontology of air. The use of Hales's pneumatic trough had already led to the isolation of a variety of chemically distinct gases, but these were considered to be ordinary air contaminated with impurities. Black was the first one to produce, isolate, and recognize a gas (carbon dioxide) as an independent substance, using Hales's term "fixed air" as the name for it. The term was ambiguous but this was ultimately irrelevant since its meaning did not determine its referent: fixed air referred to the aeriform fluid produced when magnesia or limestone (our magnesium carbonate and calcium carbonate) were burned to a calx. This new gaseous substance was rapidly incorporated into the chemical domain thanks to Black's efforts to display its relations with the better known portion of it: neutral salts. He used the analysis-synthesis cycle to break down limestone into its components (quicklime and fixed air) and then put it back together by reacting quicklime with mild alkali (potassium carbonate). The analysis and synthesis reactions could be expressed like this:

Limestone = quicklime + fixed air.
Quicklime + mild alkali = limestone + caustic alkali.

From these equations Black was able to conclude that mild alkali was composed of caustic alkali and fixed air, and this in turn implied that the increase in causticity was produced by fixed air. Since the presence or absence of the air changed the chemical behavior of the alkali, the air itself had to be a distinct chemical substance.[54] Convinced by this line of argument, Black introduced an extra column into the affinity table for this new substance.

But a more dramatic improvement of the table was performed by his Swedish contemporary, Torbern Bergman. He created a new version in 1775 with 25 acids, 16 metallic calces, and 15 earths, as compared with the original table's four acids, two alkalis, and nine metals.[55] The table was not only larger, but the inclusion of calces rather than metals made it more consistent: the table was supposed to include secondary principles, not compounds, as the metals were supposed to be. Among the novel substances included were the organic acids (tartaric, lactic, uric, citric), isolated in the 1770s and 1780s by various Swedish chemists, as well as the new "airs" isolated and analyzed in England.[56] But Bergman did more than just keep up with, and tabulate, the work of others. To answer the criticisms that had been accumulating against the original table he had to approach chemical reactions in a new way. Up to the middle of the century, chemical transformations were treated as tools to perform analysis or synthesis. A particular reaction was considered interesting if the substances it allowed to decompose or recompose were worthy of attention. In order to extend affinity research, Bergman had to disregard what was taken to be an interesting reaction and deal with all possible reactions. Through his work a *chemical transformation ceased to be a mere instrument and became a phenomenon*, one that was as much a part of the chemical domain as substances were.[57] This research led to the consolidation of Boerhaave's variant of the Affinity schema, using statement A-5":

A-5". Two substances have affinity for each other if there exists a force of attraction between them, and if the force is strong enough to overcome any existing forces of repulsion.

Bergman considered only two forces, affinity and heat. The former he conceived as a special kind of attractive force (*elective* attraction), while the latter was thought of as the source of repulsive forces. The two forces interact because as temperature rises it increases a substance's volatility and decreases the *intimate contact* needed for affinity to operate.[58] In addition, Bergman's work suggested the existence of other factors. In the course of the many experiments he had to perform to extend the table, he noticed that when two substances combined, the amount of one that it took to saturate the other was sometimes less than the amount needed to carry out the reaction. This pointed to the role played by *concentration* in chemical reactions, but Bergman did not follow this lead to its final consequences.[59] It would be left to others to venture down the road he had opened.

Affinity also fascinated Bergman's French contemporary, Louis-Bernard Guyton de Morveau. Accepting the definition of affinity as an attractive force, Guyton attempted a variety of ways to *quantify* it. One possibility was to begin with the qualitative relations already known, and use one of the chemical reactions involved as a reference point. He could then assign an arbitrary number to it, and adjust the numbers given to the other reactions.[60] Another possibility was to think of the force of affinity by analogy with other forces, such as adhesion. This led to the attempt to quantify affinity by measuring the weights that would be needed to separate different metallic surfaces from liquid mercury in contact with them.[61]

In the end, none of these methods was successful. But other aspects of Guyton's personal practice did have a lasting impact. In collaboration with Bergman, he set out to *reform chemical nomenclature*.[62] Prior to this reform, the names for chemical substances were coined following a variety of criteria: the sensible properties of a substance (its color, smell, taste, consistency); its method of preparation; the name of its discoverer or the place where it was discovered; or even symbolic associations, like that between metals and the planets.[63] The new approach was to develop a *general method for coining names* based on a substance's composition. Neutral salts were the part of the domain for which the largest number of compositional problems had been solved, so Guyton used them as his exemplary case: he created a table the first row of which contained all known acids, the first column all known bases, the resulting neutral salt placed at their intersection. Listing 18 acids and 24 bases, he could use his table to derive names for almost 500 salts.[64] Guyton's second lasting contribution was to conduct a methodical series of quantitative experiments, the results of which were used to pose the problem that would command the attention of most late eighteenth-century chemists: Why are the calces of metals heavier, rather than lighter, than the metals they come from?

Guyton's work not only established the generality of the problem for all metals—prior to him it was accepted only for lead and tin—but he linked calcination to other kinds of combustion and to acidification, conceived as a kind of "humid calcination." In this way he was able to define *a set of interconnected problems*, involving many of the known substances, thereby providing a framework for the famous controversy that followed.[65]

Reaching a solution to this set of problems involved arriving not only at a mechanism to explain the problematic gain in weight, but also working out the ontology of air. Black's isolation of carbon dioxide was the first step in this direction, but it was through the efforts of the British chemists Henry Cavendish and Joseph Priestley that the chemical study of airs (pneumatic chemistry) was developed. Hales's original apparatus used water to trap gases, but this limited its application to those gases that would not dissolve in that liquid. To capture other gases, Cavendish replaced the water by mercury, an innovation perfected by Priestley. In 1766, using this version of the trough, Cavendish isolated inflammable air (hydrogen), while Priestley went on to isolate 20 different "airs" between 1770 and 1800, including phlogisticated and dephlogisticated air (nitrogen and oxygen).[66] Despite their strange names, we know that the gases isolated by these chemists were the same as the substances to which we refer by different terms, because the reaction Cavendish used (dissolving iron in dilute sulphuric acid) does indeed produce hydrogen, and the one used by Priestley (reducing mercury calx without charcoal) produces oxygen.[67] In other words, we know what they were referring to because the referent of their concepts was fixed instrumentally. The names they gave to the gases, on the other hand, do indicate their stance towards two standing problems:

Given that metals are compounds of a calx and phlogiston, why are their calces dusty and matte, rather than ductile and shiny?
Given that the air in which a reaction occurs plays only an enabling role, why does calcination reach a definite end, rather than continuing until a metallic substance is fully transformed?

The consensus solution to the first problem was that possession of phlogiston caused metallicity, so its loss during combustion accounted for the calces' lack of ductility and sheen. Priestley not only accepted this solution but thought that the second problem could be solved along the same lines: the surrounding air becomes *saturated* with the phlogiston emanating from the burning metal, and as it becomes incapable of any further absorption the reaction stops. This solution also implied that an air that could absorb a larger amount of phlogiston, a dephlogisticated air, would be better at enabling combustion. But there was a cognitive cost in accepting these solutions: a loss of phlogiston should lead to a loss of weight in the calces, not to a gain in weight, as Guyton's results implied. Priestley and Cavendish had to deal with this discrepancy, but they could at last stop worrying about the conflict between the consensus definition of elementary substances and the fact that no one had been able to isolate and purify phlogiston. The reason why is that they thought they had finally isolated phlogiston in the form of inflammable air. But just as one of the ancient principles had been redefined, other principles came under attack. In particular, consensus practice included the belief that water was an elementary substance, but in 1784 Cavendish showed that it could be synthesized when a mixture of hydrogen and oxygen was exploded with an electric spark. Because this discovery also incurred cognitive costs, he refused to accept it, but he is nevertheless credited with it.[68]

To a younger generation of practitioners, on the other hand, it was clear that analytical contingency had caught up with Water—water, as it turned out, was not an element—just as it was catching up with Air. One of these younger chemists, Antoine-Laurent de Lavoisier, had begun his career as a follower of Hales, researching the distinction between ordinary fluid air and the air that has been fixed in solid bodies. By 1773, he had arrived at the concept of the *gas state*: all substances can become fluid and elastic like air if their temperature is high enough. He conjectured that in planets with much higher temperatures than Earth, substances that appear liquid to us would be gaseous, and vice versa, in colder planets our gases would be liquid or solid.[69] To explain these states he adopted Rouelle's concept of phlogiston as the matter of fire: a gas is just a liquid or solid to which phlogiston has been added. At this point, Lavoisier still accepted the consensus solution to the weight-gain problem, but he had changed phlogiston from a component of metals to a component of air, a displacement that would eventually lead to a novel proposal: the reason metallic calces are heavier than solid metals is because *part of the air in the reaction has become fixed in them*.[70]

Lavoisier spent most of that year as a follower of Black, carefully repeating the Scottish chemist's experiments and acquiring the skills necessary to perform pneumatic chemistry as it was practiced on the other side of the Channel. Lavoisier carefully weighed every substance that went into a reaction, as well as any substances produced by it, and whenever he could not weigh a substance directly, as was the case with carbon dioxide, he calculated their weight by measuring their volume and making assumptions about their density. His approach came to be known as "the balance sheet" approach. In retrospect, it is clear that most of the experiments conducted in 1773 were inconclusive, and that he never quite managed to balance inputs and outputs exactly. But the basic idea, that it was possible to *track the transfer* of an elastic fluid from one substance to another by relying exclusively on *conserved quantities*, was sound and it would have lasting consequences for chemistry.[71] It was at this point that Lavoisier learned about the new "airs" that Cavendish and Priestley had isolated. In 1772 he had been dealing with atmospheric air as if it was a single substance, but some of his experiments had led him to suspect that there were other elastic fluids dissolved in it. Now, the experts in pneumatic chemistry had confirmed that suspicion. By 1777, he was ready to give a more precise solution to Guyton's problem: what increased the weight of metallic calces was not the absorption of air, but of a particular component of it. This component substance was the gas that Priestley had isolated, a gas that Lavoisier would soon baptize as "oxygen." To crown his achievement, Lavoisier went on to show how his explanation of calcination as *an oxidation process* could be extended to combustion and acidification, that is, to the entire set of problems that Guyton had identified.[72]

Although Lavoisier used the concept of affinity in his account of these processes, he never showed any interest in the mechanisms behind chemical reactions. His quantitative method treated a chemical reaction as an instrument: only the conserved quantities (weights) at the start and end of the reaction mattered. But several of his contemporaries did treat chemical reactions as problematic phenomena that demanded solutions. Their goal, following in the footsteps of Guyton, was to make statements about affinities quantitative. Starting in 1781, the Irish chemist Richard Kirwan performed an extensive set of quantitative studies, using acid gases instead of liquids to increase the resolution of the measurements, and covering most of the substances in the newest affinity tables.[73] The basic idea behind his experiments was as old as Homberg: when an acid and an alkali combine to form a neutral salt, their affinity for each other should be *maximally satisfied* at the point of saturation. Thus, Kirwan argued, the amount of vegetable alkali needed to saturate a given amount of mineral acid should be considered a measure of the strength of their affinity.[74] Kirwan also extended the concept of affinity to cover more complex cases, such as chemical reactions in which two compounds are decomposed and their components switch places to form two different substances. This involved making a distinction between the affinities holding together the parts of the original compounds from those that the parts of one have for the other's, and conceiving of the outcome as the result of a contest between opposing tendencies.[75]

Research on affinities was conducted simultaneously by a young French chemist named Antoine Fourcroy. Those chemists who favored analysis over synthesis tended to focus on compositional problems and were the last ones to abandon the old notion of a principle. Priestley and Lavoisier belonged to that group: the novelty of the words "oxygen" and "hydrogen" tended to obscure their etymology, according to which they were the acid-forming and water-forming *principles*.[76] Fourcroy, on the other hand, favored synthesis, so his personal practice gravitated around the affinity forces that determined how substances combine to produce new compounds. For this reason, Fourcroy was the first chemist to assert unambiguously that *a compound has properties and dispositions* that are not present in its parts.[77] We can capture the change in his reasoning by replacing statement PW-2 in the Part-to-Whole schema, the statement asserting that the properties of a compound are a *blend* of those of its components. Another change in the reasoning patterns of chemists at the time was the realization that defining elementary substances as the limit of chemical analysis meant that the number of elements could be greatly multiplied. For instance, Lavoisier's list of substances that have not yet been decomposed included 33 different entries. This change can be captured by replacing statement PW-4, asserting that the number of basic elements or principles must be small.[78] To these two changes we must add another stemming from Lavoisier's quantitative studies: that not only the chemical nature of the components, but also their relative amounts, entered into the explanation of a compound's properties. This can be represented by replacing statement PW-5, so that by the 1780s the Part-to-Whole schema looked something like this:

Question
Why does substance X have properties Y and Z instead of other properties?

Answer
PW-1. Properties are explained by the Part-to-Whole relationship between a substance and its components.

PW-2'. The properties of a whole are a blend of the properties of its components in the case of mixtures, or a set of properties different from those of its components in the case of compounds.

PW-3. The most basic components are those that cannot be further decomposed using existing chemical operations.

PW-4'. The number of basic components is determined empirically.

PW-5'. The properties of a whole are determined by the chemical nature of its components as well as by their relative amounts.

The Affinity schema also has to be changed to reflect changes in the explanatory strategies deployed by Fourcroy. Like Macquer before him, he distinguished between kinds of affinity: the tendency to adhere displayed by substances of the same kind, like the fusing together of two drops of water, is different from the tendency of substances of different kinds to join into a new substance. This affinity of combination could be further subdivided depending on whether the chemical reaction involved a decomposition of the interacting substances, and on how many of these decompositions were involved. On this point, Fourcroy acknowledged the ideas of Kirwan on the need to take into account both the forces holding together the components of the original substances, as well as the forces tending to break them apart. These changes can be represented in the schema by adding statement A-5" and a more detailed version of A-3:

Question
Why does the reaction of substance X with substance Y have substance Z as its product, instead of other products?

Answer
A-1. Substances possess dispositions to selectively combine with other substances.

A-2. If X and Y are elementary substances then their dispositions will determine Z: if they have affinity for each other their product will be a combination, else they will not combine.

A-3. If X and Y are compound substances then the dispositions of their components will determine Z:

> A-3a. If only one of the substances is a compound, then the elementary substance's affinity for one of the compound's components will cause it to be dislodged and join with it, displacing the other component.

> A-3b. If both substances are compounds, then they will both decompose and their components will unite following their respective affinities.[79]

A-4. The affinity of a substance for another substance is constant.

A-5". Two substances have affinity for each other if there exists a force of attraction between them, and if the force is strong enough to overcome any existing forces counteracting attraction.

It was as part of the personal practice of another French chemist, Claude-Louis Berthollet, that the consensus belief in statement A-4 was finally challenged. Berthollet did not question the existence of affinities, but he was bothered by the fact that statements about them were made solely on the basis of the simplest displacement reactions, that is, those involving an interaction between three substances. In this case, it was easy to assume that if X has a higher affinity for Y than for Z, then when X interacts with a compound of Y and Z it will *completely displace* Z.[80] But reactions involving more than three substances could conceivably have more complex outcomes. Moreover, Berthollet had observed that in industrial processes the tendency to combine decreased in proportion to *the degree of combination already achieved*. In other words, saturation effects could affect the constancy of affinity, and these effects would be more significant in reactions involving more than three substances: their proportions, and more significantly, their degree of concentration, were as important as their affinities in determining the final outcome.[81] Thus, he concluded that displacement reactions were poor exemplars to guide future research on affinity, and recommended switching to *neutralization reactions*, like those used by Kirwan to measure affinity.[82]

Besides introducing concentration as a factor that can affect and even reverse the order of affinities, Berthollet stressed the role played by temperature, drawing on the lessons provided by other chemists working on the ontology of heat. Specifically, Lavoisier's first step in his solution to the weight-gain problem had been to displace the matter of heat from the interior of metals to the air around them. He eventually renamed this substance "caloric," to stress its differences with phlogiston: the fact that it emanated from all substances, not only those that were inflammable, and that it could be made part of the balance sheet by measuring it with a thermometer.[83] Although the concept of heat as a substance was faulty and was eventually replaced, it played an important role in accounts of the volatility and fluidity of substances, and as such it could be used to show how these factors affect affinity: if, as Fourcroy and others thought, the forces of attraction involved were different than gravity in that they demanded *intimate contact* between interacting particles, then any factor that prevented such a contact had the potential to change the constancy of affinity relations.[84] If a substance became gaseous in the course of a reaction, for example, it would be too volatile to maintain this intimate contact. These changes in Berthollet's reasoning can be captured if we replace statement A-4 by A-4':

A-4'. The affinity of a substance for another substance is not constant but can change depending on the effect of a variety of other factors: concentration, temperature, solubility, elasticity, efflorescence.

Let's move on to consider the consensus practices that had emerged by the year 1800. This consensus has been traditionally portrayed as the result of the revolutionary overthrow of the irrational "phlogiston theory" by the rational "oxygen theory," or conversely, as the clash between two incommensurable world-views neither one of which can be objectively judged to be better than the other. But as we have already shown, if we replace monolithic theories with a multiplicity of cognitive tools it is clear that there was as much continuity in this controversy as there was discontinuity. All participants used reasoning patterns that conformed to some version of the Part-to-Whole schema: their disagreements were over whether a substance was an element or a compound, not over the validity of explanations in terms of parts and wholes. Similarly, both sides accepted explanations in terms of dispositions to combine, that is, both used reasoning patterns that conformed to the Affinity schema. If these partial overlaps are ignored it becomes impossible to see how the chemists involved could have been convinced to switch sides through cogent argumentation and judicious assessment of evidence. Thus, when Berthollet accepted Lavoisier's arguments in 1785 and changed his mind about phlogiston, this was not the result of a reasoned evaluation of the merits of the respective proposals but something more like a religious conversion. And similarly for Fourcroy in 1786, Guyton in 1787, and Kirwan in 1791.

But because there were shared concepts, statements, problems, and taxonomic and explanatory schemas, a *collective argument* could emerge from the debate. When Lavoisier was the only defender of the oxygen model of combustion he formed a discussion group (the Arsenal group) to which Fourcroy and Berthollet belonged. It was within this group that the oxygen model was further developed and extended to acidification, as part of *a community of opinions*, in Lavoisier's own words, in which the role of affinities and heat in chemical reactions served as a common thread.[85] Hence, when Berthollet switched sides after a demonstration of the synthesis of water, it was not because of a sudden transformation of his world-view, but as a consequence of the comparison he could now make between his own ideas about acids and the new concepts discussed in the group. Guyton took longer to abandon phlogiston, but when he did it was because he realized how close his ideas were to Lavoisier's own conception of heat.[86] Priestley and Cavendish were never convinced, but this was not because they lived in a different world. In fact, they shared more with Lavoisier than either side would have admitted. In particular, all three shared a view of composition in terms of principles, as opposed to one based on affinity: when Lavoisier presented his list of elementary substances, he grouped together light, caloric, oxygen, hydrogen, and nitrogen as the most basic, a grouping that reveals he was still thinking in terms of a few fundamental principles, much like his rivals.[87]

Nevertheless, because in the eyes of contemporary chemists the French side won the controversy, it will be safer to assess the consensus of 1800 by using a Scottish textbook.[88] Conceptual improvement can be immediately noticed by looking at the ontology of the incorporeals: both air and heat are better conceptualized than they were in 1750. Air is not a basic principle anymore, and is now thought of in terms of the *gas state*, a state common to all substances. The concept of a gas, however, is still underdeveloped because heat is still conceived as a chemical substance, which erroneously makes any gaseous substance a compound.[89] Despite this shortcoming, the extensive research of the previous 50 years on the production of heat in combustion and vaporization has added important information that would eventually improve the concept, such as the fact that substances have an excess of heat when in gas form, an excess that remains *latent* until the heat is released when the gas liquefies. This led chemists to make a conceptual distinction between the amount of heat and the intensity of heat, the latter being the property that thermometers measure. Another new concept that has been added to the consensus is *specific heat*: the amount needed to change a substance's temperature by a given amount. By measuring the specific heat of oxygen gas and other combustible substances at the start of the reaction, and the specific heat of its products at the end, for example, it could be shown that changes in this property occurred as a result of the transformation, an important step in the elucidation of the chemical role played by heat.[90]

Because specific heat varies from one substance to another, it could be used as a means to identify them, that is, specific heat became another way to fix the referent of substance concepts. Other property concepts, like *specific gravity*, also played this identifying role. This concept refers to a substance's comparative density, a property measured as the ratio of the substance's density to that of another substance used as a standard, typically water at a temperature of 4°C. If a substance's specific gravity is lower than another's, it floats in it; if it is higher, it sinks. The property itself was not unknown—the phenomenon was the basis for instruments used in breweries and distilleries, as well as of techniques like panning, jigging, or shaking used in mines to concentrate ores—and a word for it had existed for decades.[91] But it was not until the advent of pneumatic chemistry that specific gravity became an identifying trait necessary for distinguishing one gaseous substance from another. Thus, while in the textbook of 1750 specific gravity was used only a few times, by 1800 it has become an identifying property of every substance.[92]

The concept of a compositional hierarchy, already part of the mid-century consensus, was improved by making explicit the notion of an *irreducible property*. In the study of plant and animal substances it was generally accepted that carrying chemical analysis to its limit often led to products that had lost their medicinal virtue, so that analysis should stop with more proximate principles.[93] This piece of practical advice implied that "medicinal virtue" was a property that emerged at a certain level of the compositional hierarchy, and that it ceased to exist at lower levels. Making this idea explicit, however, involved making a conceptual distinction between mixtures, with properties that are a blend of those of its ingredients, and compounds with entirely novel properties. In 1800 the taste, smell, color, form, density, fusibility, volatility, solubility, and other properties and dispositions of a compound are characterized as being entirely different from those of its components. Affinity itself is now defined as a form of contiguous attraction that, unlike cohesion, generates substances with properties totally dissimilar from those of its components.[94] This indicates that the conceptual change was brought about by studying composition not in terms of principles but in terms of the process that generates chemical union.

On the other hand, although mixtures and compounds can be told apart conceptually, the distinction has not yet been made instrumental: a pure compound cannot yet be generally distinguished from a substance that may contain a mixture of related compounds.[95] This limitation was made more evident by the recent discovery that factors like temperature and concentration affected the outcome of chemical reactions, and that one and the same reaction could produce a variety of mixed products.

Nevertheless the concept of a *pure substance*, elementary or compound, has become consolidated in the new consensus. The concept had existed implicitly for a long time as part of purification procedures for precious metals, but it had to be generalized. In the 1750 textbook the importance of purity is emphasized, but mostly in relation to substances that had concerned assayers for centuries.[96] In 1800, purification procedures are discussed as part of the very characterization of the identity of almost every elementary or compound substance.[97]

Among the many statements that have joined consensus practice by 1800, some were the product of the massive reassignment of substances to the categories of elementary and compound. Statements like 1 and 2 have been changed to 1' and 2', to reflect new beliefs about composition:

1 Metallic calces are elementary substances.

2 Metals are compound substances.

1' Metallic oxides are compound substances.

2' Metals are elementary substances.[98]

There are also many new statements giving the composition of substances previously thought to be elementary, statements like:

3 Ammonia is a compound of nitrogen and hydrogen.[99]

Another source of new statements was the addition of quantitative information about composition. The relative quantities of components could be expressed in a variety of ways, as percentages or as parts per weight:

4 Ammonia is composed of 121 parts per weight of nitrogen and 29 parts per weight of hydrogen.

5 Pure water is composed of oxygen and hydrogen in proportions 85 to 13.

6 Nitrous oxide is composed of nitrogen and oxygen in a 63:37 ratio.

7 Nitric oxide is composed of nitrogen and oxygen in a 44:56 ratio.

Using induction, a powerful empirical generalization could be formulated from many statements like these, a generalization expressed by a statement like:

8 Two substances combine in either a single definite proportion (oxygen and hydrogen in water) or in proportions that are multiples of one another (oxygen and nitrogen in nitric and nitrous oxide).[100]

Many novel problems had been added to the consensus practice of chemists by 1800. The very introduction of new concepts and statements made certain phenomena problematic. Thus, adding heat capacity as an identifying property of substances raised the question: Why do different substances at the same temperature contain different amounts of caloric, rather than the same amount?[101]

Similarly, the introduction of statements about the composition of the oxides of nitrogen and of atmospheric air, raised questions like: Why do atmospheric oxygen and nitrogen exist as a mixture, rather than as a compound (as in nitrogen oxides)?[102]

Acknowledging the existence of complex affinities raised questions about their mechanisms. Even simple affinities could present chemists with challenges, such as explaining the lack of transitivity displayed by some affinity relations. These problems can be easier to understand if we list their presuppositions:

Given that substance X displaces Y in the compound YZ to form the new compound XZ, then why does Y displace X from XZ to form YZ, rather than not displacing it at all?[103]

Given that substance X completely displaces Y, and Y completely displaces Z, then why does X does not displace Z (or displaces it only incompletely) instead of displacing it completely?[104]

Existing reasoning styles were greatly improved in this period. The form they had acquired by 1800 can be represented by the changes in the explanatory schemas discussed above.[105] But new explanatory devices have also been introduced, like the diagrams created by Bergman and others to tackle complex affinities.[106] Classification schemes have been made better through extension, but these more comprehensive tables of affinity have probably reached a limit of improvability. The reason for this was the recognition of the role of temperature, concentration, and other factors that can affect affinity relations. Strictly speaking, adding the effects of these other factors would have involved creating a separate table for every degree of temperature or concentration, greatly reducing their taxonomic value.[107] But there were also taxonomic novelties, the main example of which is the *system of nomenclature* of Bergman and Guyton, the final form of which included important contributions from Fourcroy and Lavoisier. As discussed above, the original project was not merely a matter of improving the language used by chemists by giving new names to old substances.[108] The goal was to develop a method for *generating* new labels for existing substances, as well as for those that would be separated and purified in the future. The very fact that a table was used to generate the names shows that the main goal of the system of nomenclature was to bring order to an ever growing domain. Joining the new names to the old compositional hierarchy could accomplish this task: the base of the hierarchy had to be explicitly defined as composed of substances that were the limit of chemical analysis: gaseous elements like oxygen, hydrogen, and nitrogen; solid elements like gold, iron, and copper; as well as elements that combined with oxygen to form acids, like sulphur, phosphorous, and carbon. Once the base of the hierarchy was laid down, the other layers could be added and names derived from those of elementary substances: oxides and acids; the alkalis and other bases of neutral salts; minerals; vegetable and animal compounds.[109]

2 ORGANIC CHEMISTRY

The Specialization of Cognitive Tools

So far we have examined how a variety of cognitive tools can be improved over time, and how the existing set can be enriched through innovation. But there is another type of change that a body of cognitive products can undergo: differentiation or specialization. The source of this kind of change is associated with the splitting of a single field into several subfields, as when classical chemistry in the nineteenth century diverged into inorganic and organic chemistry. In this chapter we will explore some of the changes caused by the extension of the practices that had developed around neutral salts to the portion of the chemical domain that included plant and animal substances. Because organic reactions are typically very complex, often resulting in a cascade of products difficult to analyze quantitatively, the extension of the lessons of the previous century was often based on analogies and conventions. These formed, as it were, a temporary scaffolding that could be discarded once the bridge between the two parts of the domain had been built.

Quantitative statements about composition were the first cognitive tool that had to be changed. The balance-sheet method developed in the study of combustion—carrying out reactions in a closed vessel while carefully weighing the substances present at the start and at the end of the transformation—yielded quantities measured in conventional units of weight. But these absolute quantities were less useful when applied to organic substances composed of only a few different elements: carbon, oxygen, hydrogen, and nitrogen. A better approach to characterize these compounds involved using *pure numbers* designating proportions, rather than absolute numbers expressed in units of weight. This way the large variety of organic substances could be explained as the result of changes in the relative amounts of a few building blocks. Important steps in this direction had already been taken in the last decade of the eighteenth century, as part of an effort to calculate the relative proportions of acid and base ingredients in a neutral salt. This method involved measuring the amount of a particular acid needed to neutralize a given amount of a particular base, or more exactly, to measure the *number of parts per weight* of an acid needed for the neutralization of a number of parts per weight of a base. Once this determination had been made for one acid, several bases could be compared with each other by calculating the amounts needed to neutralize the same reference acid. Thus, if 1000 parts per weight of sulfuric acid was used as a standard, the number of neutralizing parts of different bases could be determined: 525 parts of aluminum oxide, 615 parts of magnesium oxide, or 672 parts of ammonia. These numbers without units were called "equivalent weights," or simply, *equivalents*.[1]

A rival approach to the determination of pure numbers was based on the *atomic weights* of substances in the gas state. To determine the atomic weight of carbon, for example, a diamond was exposed at white heat to a current of purified oxygen, which reacted with the crystalline carbon to produce a gas (carbon dioxide). Comparing the weight of the original diamond with that of the carbon dioxide, the ratio of carbon to oxygen could be determined. Thus, the original experiments showed that for every 3 parts per weight of diamond, there were 11 parts of carbon dioxide. Using carbon as a standard substance, and arbitrarily assigning it the reference weight of 12, carbon dioxide would get the number 44. Finally, using old balance-sheet information about this gas it could be concluded that the atomic weight of oxygen was 16.[2] It is important to emphasize that despite the use of the word "atomic," these pure numbers represented conventional comparative weights. If a chemist had the need to visualize what these pure numbers referred to, this could be achieved by picturing extremely small *portions* of a macroscopic substance. Thus, equivalents and atomic weights could be treated as *units of composition*, in the sense in which we speak of atoms composing a molecule, or as *units of combination*.[3] In what follows, the term "units of combination" will be used because most nineteenth-century chemists could accept the validity of these numbers without having to commit themselves to the existence of atoms.

The need to use an arbitrary substance as a fixed point of reference for both equivalents and atomic weights meant that, in the absence of a commonly accepted convention, units of combination varied from one chemist to another. Moreover, practitioners had plenty of discretion to round up the messy figures from laboratory measurements into integer numbers, providing another source of variation.[4] But underneath the variations there were regularities that the conventionality of the numbers could not hide. In particular, the combining proportions tended to occur in simple multiples: oxygen and nitrogen, for instance, were able to form compounds in proportions 37 to 63, 56 to 44, or 70 to 29, but not in any other proportion. This led to the important empirical generalization that two substances could combine in either one determinate proportion or in several proportions that were integer multiples of one another.[5] In the early 1800s, not everyone accepted this generalization, and the question of whether proportions varied continuously or discontinuously could not be resolved experimentally.[6] But there was an advantage to accepting the validity of the general statement, because *discontinuous variation in proportions implied definite compositional identity*. For instance, the difference between two kindred substances like nitric oxide and nitrous oxide could be determined unambiguously if chemists could assume that substances with intermediate combining proportions did not exist. The techniques developed around combinatorial units, collectively known as "stoichiometry," helped organic chemists demarcate the boundaries of their domain, as they selected from the confusing mass of resins and oils, fibers and gums, only substances with a *well-defined stoichiometric identity*.[7]

Quantitative statements about composition using the new units of combination were not the only cognitive tool used to clarify the identity of the organic domain. There were also new taxonomic schemas giving order to it. We argued in the previous chapter that the new nomenclature introduced in the late eighteenth century was meant to replace not only old labels with new ones, but to serve as means to classify all substances by their composition. A similar point can be made about the *chemical formulas* first introduced in the second decade of the nineteenth century. These formulas used letters as symbols for elementary substances, the first letter of their Latin names, but this was not in itself an innovation since chemists had used letters before.[8] The most important new feature was that *a letter stood for a unit of combination*, and was accompanied by a numeral when the number of units was greater than one. Joining two letters with a plus sign (and later on, juxtaposing them) signified chemical union into a compound, not mere coexistence in a mixture. The formulas took decades to acquire a stable form: water, for example, was represented variously as $2H+O$, H^2O, H_2O, H^4O^2, and HO. Underneath the mass of confusing variations, however, practitioners rightly saw a tool for recognizing relations between substances, and hence, a way to relate them to one another taxonomically.[9]

To bridge the gap between the classification of mineral substances and that of vegetable and animal ones, conventions for units and formulas had to be supplemented by analogies. Inorganic taxonomies were based on the assumption that all compounds were *binary*: neutral salts were always made of an acid and a base and, with the exception of metals, acids and bases themselves were binary. Many acids, for example, were composed of oxygen together with another component that varied from one substance to the next: carbon, sulphur, phosphorous. The unvarying component was referred to as a *radical*.[10] This idea could be extended by analogy to *groups of elements* in organic substances that retained their identity during transformations. In the 1820s and 1830s many of these groups were postulated to exist, and given the name of "organic radicals." Their identity was not instrumental, since their existence was inferred from the manipulation of formulas, but they were believed to exist and were given distinctive names, like ethyl (C_2H_5), methyl (CH_3), phenyl (C_6H_5), benzoyl (C_6H_5CO), and amino (NH_2).[11] Organic radicals were a recurring feature of the formulas of substances obtained in the laboratory by successive transformations of a parent substance, such as alcohol or ether. This generated *series of substances* in which something remained constant while something else varied, and these series could be overlaid over the jungle of plant and animals extracts to reveal hidden kinships. This way, chemical formulas began to change from a purely empirical record of proportions to a model of composition in which functional groups were highlighted. Thus, the empirical formula for a substance like the oil of bitter almonds, C_7H_6O, became the *rational formula* (C_6H_5CO) + H, in which the radical benzoyl was highlighted.[12]

The only problem with the analogy between organic radicals and elements as basic building blocks was that while the referent of terms for elementary substances was fixed instrumentally, the referent of terms for radicals could not be fixed that way because they resisted isolation. The only evidence for their autonomous existence was the length of the series of derivatives in which they figured as invariant components.[13] Another limitation of the radical approach was that the chemical reactions used to generate series were limited to addition and elimination, reactions that were analogous to the displacement reactions of the eighteenth century.

In these reactions *like always displaced like*: a metal dissolved in acid was displaced by another metal, and an alkali displaced another alkali in a neutral salt. Moreover, the spread of electrochemistry after 1800 fostered the idea that a replacement could occur only among similarly charged components: electropositive elements or radicals were assumed to be replaceable exclusively by other electropositive elements or radicals.[14] An alternative classification scheme emerged when chemists began using a new type of chemical transformation, *substitution reactions*, in which an elementary component was replaced by a *different* one.

The radical approach to classification, and its associated binary model of composition, was challenged in 1834 when a reaction was found in which the substitution occurred between elementary substances with *opposite* electrical charges: hydrochloric acid was produced by replacing one unit of electropositive hydrogen by one of electronegative chlorine. Soon other reactions were discovered in which hydrogen units were replaced by units of electronegative elements like bromine and iodine.[15] According to the radical approach these substitutions should be impossible, so their discovery led to the development of a different way of classifying organic substances. This was also a serial taxonomy, based on sequences of chemical reactions in which constants were identified and variations multiplied, but the constants in this case were what was called *types*. In the type approach, the formula for water, for instance, was written like this:

$$\left.\begin{array}{c} H \\ \\ H \end{array}\right\} O$$

This format allowed the similarity of water with substances like ether or alcohol to be grasped at a glance, if their respective formulas were rendered like this:

$$\left.\begin{array}{c} C_2H_5 \\ \\ C_2H_5 \end{array}\right\} O \qquad\qquad \left.\begin{array}{c} C_2H_5 \\ \\ H \end{array}\right\} O$$

By the middle of the century there were four different types—the water, ammonia, hydrogen chloride, and hydrogen types—allowing the classification of a large number of organic substances. Thus, all oxides, sulfides, salts, alcohols, and ethers—substances that were susceptible to the same substitution changes as water—were made members of the water type; all amines, nitrides, phosphides, and arsenides were grouped under the ammonia type; all halides and cyanides were gathered under the hydrogen chloride type; and paraffins, metals, and metal hydrides under the hydrogen type.[16] With some exceptions, chemists did not treat type formulas as *iconic*. The symbol for oxygen in the formula for ether, for example, was not interpreted as meaning that an oxygen atom served as the link (or bond) between two ethyl radicals in a molecule. Rather, the symbols were arranged to show what combinatorial units were more likely to be replaced in a substitution reaction.[17] In other words, most chemists did not regard these formulas as representations of an invisible reality but as *memoranda of the visible reaction history of a substance*.[18] Nevertheless, if we ignore this for the moment, the difference between the radical and the type approaches to classification was that the former focused on groups of "atoms" while the latter was concerned with a "molecule" as a whole.[19]

To summarize what has been said so far, as the old cognitive tools were adapted to organic chemistry, new concepts were introduced to express combinatorial proportions (equivalents, atomic weights); new formulas introduced to model a substance's composition in relation to the transformations it was capable of undergoing; and new taxonomies created that were serial rather than tabular. Problems and the explanatory schemas used to solve them were also affected by specialization. In the Part-to-Whole schema, these changes were at first relatively minor. As we saw in the previous chapter, the last statement in this schema had changed from PW-4 to PW-4':

PW-4. The most basic components are few in number.
PW-4'. The number of basic components is determined empirically.

The change was necessary to accommodate the increasing number of substances that could not be further decomposed by chemical means: from the 33 with which the eighteenth century closed, to the 54 featured in textbooks of the 1830s, to the 70 elementary substances in the Periodic Table of the late 1860s.[20] Another change in chemists' reasoning patterns, a change already part of the consensus of 1800, can be represented by the replacement of statement PW-5 by a new one stating that the properties of a whole are determined both by the nature of its components as well as by their relative proportions, that is, statement PW-5'. But the problem of the singular diversity of organic compounds arising from a repertoire of only four basic building blocks seemed to require further changes. In particular, by mid-century some chemists began to wonder whether the *spatial arrangement* of building blocks was also a causal factor, a change in their explanatory strategies that can be captured by statement PW-5":

PW-5". The properties of a whole are determined by the nature of its components, by their relative proportions, as well as by their spatial arrangement.

Chemists with a background in crystallography had no problem accepting such a change. They knew that neutral salts with different metal components could have similar crystal shapes, and that some organic substances obtained from substitution reactions also exhibited this isomorphism. To them, this phenomenon conflicted with PW-5' since similar spatial properties resulted from components having a different chemical nature.[21] But even for chemists for whom crystal shape was unimportant, some substitution reactions posed a similar problem. We said above that a counter-example to the radical approach was the discovery that an electronegative element like chlorine could substitute an electropositive one like hydrogen while leaving some properties of a compound unchanged. Why? One possible explanation was to postulate that it was the *position* of the chlorine that mattered, that is, that chlorine when buried within a larger "molecule" was as electrically neutral as hydrogen.[22] Thus, whether prompted by crystal isomorphism or by puzzling substitutions, practitioners had to confront the problem: Why do substances having *different* compositions possess *similar* properties, rather than different properties?

Schemas like Part-to-Whole were not the only explanatory devices available to organic chemists. In particular, chemical formulas could be used to solve problems if instead of modeling substance composition they were used to model chemical transformations. A good illustration is a reaction that had been known to chemists and apothecaries for a long time, a reaction in which ether was produced by reacting alcohol with sulphuric acid. In 1797 the reaction had been studied and explained in terms of affinities: sulphuric acid had a greater affinity for water, so it removed the right proportions of hydrogen and oxygen from alcohol to yield ether. By the 1820s, chemists could use formula equations to obtain a similar solution. Thus, given the formulas for alcohol (C_2H_6O) and ether ($C_4H_{10}O$), practitioners could see at a glance not only that these were related substances, but that one could be produced from the other by the removal of water. The process that produced one volume of ether from two of alcohol could be modeled with the following equation:

$$C_4H_{10}O = 2 \, (C_2H_6O) - H_2O^{23}$$

Equations like these were more than just restatements of affinity arguments. Unlike inorganic reactions, organic ones produced a variety of additional products that were hard to capture and analyze. The ether-producing reaction was no exception, and additional products (sulfovinic acid) had been identified by 1820. This led to a specialization of the questions that the Affinity schema was supposed to answer, questions that now took the form: Why does reaction X produce product Y (and byproducts A, B, C) instead of product Z (and byproducts A', B', C')?

Solving problems like these involved simulating the method used in the laboratory, in which the weights of the reactants and those of the products were balanced, using formula equations. The basic assumption of the balance-sheet method, the conservation of weight, was simulated by *keeping the number of combinatorial units constant*. In a pioneering application of this idea, French chemists showed that the above reaction could be modeled as taking place simultaneously along two pathways, one producing ether by dehydration and another yielding the additional byproducts.[24]

The extension of the use of formulas from a taxonomic role to an explanatory one was also motivated by compositional problems. We just mentioned one of these, the problem of explaining why properties could be shared by compounds made of different elements or radicals. The urgency of problems like these increased when a phenomenon recently added to the domain, the phenomenon of *isomeric substances*, posed the complementary problem: Why do substances having *similar* composition possess *different* properties, rather than similar properties?[25]

Explaining why two substances could behave differently without any change in their composition motivated the creation of a new type of formula, *structural formulas*, in which in addition to the nature and relative quantities of the components, their possible *connectivity* was modeled. Displaying graphically how elements and radicals could be linked to one another fell short of expressing their spatial arrangement, but for a while this was a benefit rather than a shortcoming, since it allowed chemists with different ontological commitments (atomists and anti-atomists) to use the new formulas. As it turned out, explaining the phenomenon of isomerism would demand a stronger embrace of spatial arrangement as a causal factor, and of formulas that were more iconic.[26] Nevertheless, the earliest structural formulas could be used to answer the much simpler question: Why does substance X have Y number of isomers, rather than Z number of isomers?

The problem posed by this question became pressing in the second half of the nineteenth century as isomeric variants began to proliferate, and the *number of variants* was shown to display clear regularities. One such substance was lactic acid, an organic substance long known to have at least two isomers, one extracted from spoiled milk, the other from animal muscle. As long as only two variants were known to exist, non-iconic structural formulas could be used to account for them in terms of connectivity patterns, and of the relative position of a radical in these patterns.[27] Thus, starting with the empirical formula for lactic acid, $C_3H_6O_3$, the following two structural formulas could be used to account for its known variants in terms of the relative position of the radical OH:

$$\text{H}-\underset{\underset{\text{H}}{|}}{\overset{\overset{\text{H}}{|}}{\text{C}}}-\underset{\underset{\text{H}}{|}}{\overset{\overset{\text{OH}}{|}}{\text{C}}}-\overset{\overset{\text{O}}{\|}}{\text{C}}-\text{OH} \qquad \text{HO}-\underset{\underset{\text{H}}{|}}{\overset{\overset{\text{H}}{|}}{\text{C}}}-\underset{\underset{\text{H}}{|}}{\overset{\overset{\text{H}}{|}}{\text{C}}}-\overset{\overset{\text{O}}{\|}}{\text{C}}-\text{OH}$$

One difficulty with these structural formulas was that the depiction of the connections of carbon at 90-degree angles led to predicting the existence of a larger number of isomers than was actually known. To fix this problem a third dimension was added: the carbon connections were modeled as if they were directed towards the vertices of a tetrahedron, with the carbon at the center of the triangular pyramid connected to four different elements or radicals placed at its vertices. With this added dimension, arguments based on the symmetry of the formulas could be used to constrain the number of predicted isomers to match the number actually observed in the laboratory.[28] These three-dimensional formulas began the trend towards more iconic representations, although the icons did not attempt to represent the spatial structure of a carbon atom—a structure no one suspected any indivisible, ultimate particle could have—but carbon's *tetrahedral grasp* when the element was part of an organic compound.

To understand the rise of structural formulas we need to discuss just what this "tetrahedral grasp" was supposed to be. Ideas like these emerged when the disposition of a substance to combine with another, that is, its affinity, began to be conceived in terms of the capacity of its component elements or radicals to combine *with a specific number* of other elements or radicals. This numerically regular capacity was named the *valency* of a substance.[29] The concept of valency emerged slowly as lines of reasoning deriving from both the radical and type approaches to classification converged. On the one hand, chemists using radicals noticed that elements like nitrogen tended to form compounds containing either three or five combinatorial units of other elements, and conjectured that their affinities were best satisfied in those proportions.[30] On the other hand, chemists who practiced taxonomy using types, noticed that while one unit of chlorine replaced one of hydrogen, other elements had other hydrogen-replacing powers: tin could replace two hydrogens, while bismuth could replace three, for example. These numerical regularities could be accepted as a brute fact, in which case valency was conceived as *a mere number* that was useful for taxonomical purposes. But the valency number could also be taken as an indication that affinity, which had been traditionally conceived as a unified macroscopic force, was subdivided at the microscopic level into *affinity units*. For these chemists the combination of the affinity units of different atoms represented the formation of a chemical *bond*.[31]

Using the record of a substance's chemical reactions as a source of evidence, it could be inferred that carbon had four affinity units, oxygen had two, and hydrogen one. Given the empirical formula for a substance like acetic acid, $C_4H_4O_4$, chemists could calculate that the four carbons had 16 affinity units altogether, four used to bind the monovalent hydrogens and eight to bind the divalent oxygens. The remaining four affinity units could be explained by assuming that they were used by the carbons to bind with each other, forming *chains*. These chains were the first kind of spatial structure to enter into the solution of compositional problems. But the idea that carbon could link with itself also played a crucial role in the development of the concept of valency. All together, this concept involved four separate ideas: an element's possession of a maximum combining power; the capacity of elements or radicals to replace multiple hydrogens; the tetravalency of carbon; and the self-linking of carbon. It can be argued that these four ideas never coexisted in a single mind, vividly illustrating the collective way in which some cognitive tools are developed.[32] Structural formulas that included all four ideas, like the formula for benzene, became exemplary cases in their time. Benzene's empirical formula was C_6H_6, suggesting a carbon component with 24 affinity units, six of which were used to bind the hydrogens. The rest of the units were explained by postulating that the carbons formed chains, and that these chains were *closed* (and that every other carbon formed a double bond). The structural formula for this closed chain, a hexagonally shaped ring in which every corner contains a carbon bonded to a hydrogen, is still ubiquitous today.[33]

Let's summarize the changes that problems and the explanatory strategies used for their solution underwent as chemistry differentiated into organic and inorganic subfields. We can do that by adding to the old Affinity schema the set of new statements that the concepts of valency and bonding made possible. Statement A-1 remains the same, but it has been augmented with three new statements about the relation between affinity and valency. Statements A-2 and A-3, on the other hand, have been replaced by A-2' and A-3' respectively. The new reasoning pattern can be represented like this:

Question
Why does the reaction of substance X with substance Y have substance Z as its product, rather than other products?

Answer

A-1. Substances possess dispositions to selectively combine with other substances.

A-1a. Combinations are formed by the creation of bonds between the atoms composing those substances.

A-1b. Each bond involves the satisfaction of two affinity units, each of which has a direction in space.

A-1c. The number of affinity units, and therefore the number of possible bonds, is fixed.

A-2'. If X and Y are elementary substances then their disposition to form bonds is determined by their valency number. For example, if X is tetravalent and Y monovalent, each atom of X will form a bond with four atoms of Y.

A-3'. If X or Y are compound substances then their disposition to form bonds is determined by their valency number, which can be calculated like this:

A-3a'. If the compound is made of monovalent elements then its valency number equals the sum of the affinity units of its elements minus the affinity units used to form bonds between different elements.

A-3b'. If the compound is made of polyvalent elements then its valency number equals the sum of the affinity units of its elements minus the affinity units used to form bonds between different elements and those used to form chains or rings of the same element.[34]

A-4'. The affinity of a substance for another substance is not constant but can change depending on the effect of a variety of other factors: concentration, temperature, solubility, elasticity, efflorescence.

A-5". Two substances have affinity for each other if there exists a force of attraction between them, and if the force is strong enough to overcome any existing forces counteracting attraction.

Needless to say, only atomists could subscribe to the full version of the schema. Anti-atomists could accept statements A-2' and A-3', since these statements use the ontologically neutral phrase "valency number." They would have warned, however, that the phrase "affinity unit" in both A-3a' and A-3b' should be taken with a grain of salt. Only statements A-1a to A-1c would have been rejected outright as being too speculative. On the other hand, not all atomists accepted the changes expressed by A-1b and A-1c. In particular, inorganic compounds seemed to violate the idea that a given elementary substance had a *fixed valency*. Nitrogen, for one, seemed to have three units in some compounds and five in others.[35] A particular class of substances, compounds of ammonia and a metal, became exemplary in this regard because of their dazzling colors that changed systematically with the number of ammonias (NH_3): yellow with six ammonias, purple with five, green with four.[36] If we represent the metal (cobalt, platinum, and rhodium) by the letter M, and use the letter X for the elements other than ammonia (chlorine, bromine), then the composition of these compounds could be represented by formulas like these:

$$M\,(NH_3)_6\,X_3 \text{ (yellow)}$$
$$M\,(NH_3)_5\,X_3 \text{ (purple)}$$
$$M\,(NH_3)_4\,X_3 \text{ (green)}$$

To the chemists working on these organo-metallic compounds, the formulas implied that the metal atoms had multiple valencies, and that the way in which they were bonded with the ammonias was problematic if the idea that affinity was split into units was accepted. So they rejected that concept, conceiving of affinity as a *continuous* electrical force. The metal atom in these colorful compounds was conceived as possessing a spherical distribution of affinity around it in which the primary binding with the ammonia groups occurred, and the ammonias distributed in the sphere in such a way as to maximize bonding capacity. These ammonias, in turn, were thought to convey some of the force of affinity to a second outer sphere in which the bonds with elements like chlorine or bromine were located.[37] Clearly, the pattern of reasoning used by these chemists is not represented by the previous version of the Affinity schema. So we must create another version to fit this case: statements A-2' and A-3' have to be changed so that in addition to the valency number the *coordination number* (the number of bonds that a metal atom can form) is taken into account; and statements A-1b and A-1c have to be replaced by A-1b' and A-1c':

A-1b'. Affinity forces are distributed continuously with spherical symmetry, and as many bonds are formed in this sphere as it takes to maximize bonding capacity. A second sphere exists around the first one, serving as a locus for additional bonds.

A-1c'. The number of possible bonds is not fixed.

The existence of three different variants of the Affinity schema, two compatible with atomism and one compatible with anti-atomism, suggests that debates over their respective merits were inevitable. And a similar point applies to all the other cognitive tools discussed so far:

there were heated controversies over combinatorial units (equivalents against atomic weights); taxonomies (radicals against types); formulas (empirical against structural); and explanatory schemas (fixed valency against variable valency). Like their eighteenth-century counterparts, these debates could be harsh and bitter, and were often accompanied by priority disputes.[38] Improvement of the different cognitive tools, to the extent that it occurred, did not follow a straight line of development. On the contrary, there were many false starts, dead ends, precarious truces, compromises between positions, and authoritative practitioners holding on to ideas that the majority had already given up.[39] But at no point can we detect a complete breakdown in communication, not even between the holders of diametrically opposed ontologies. Moreover, contributions of rival approaches often became part of the consensus, regardless of what approach had been declared the winner of a controversy. Thus, despite the fact that the type approach to classification prevailed over the radical approach, both contributed to the development of the concept of valency: the type approach supplied the idea that elements or radicals could be polyvalent (a capacity for multiple bonding), while the radical approach contributed the notion that one and the same element could have variable valency (a capacity to form a different number of bonds in different combinations).[40]

Let's do a rough assessment of the improvement of cognitive tools in this century, starting with substance concepts and the means to fix their referent. In inorganic chemistry the identity of a compound was established by the nature of its components as revealed by chemical analysis, and then corroborated by synthesis. But the procedure of taking a substance apart and then putting it back together again could not be applied to organic compounds because techniques for synthesis lagged behind those for analysis for most of the century. Adding quantitative information to statements about composition mitigated this problem, but the lack of agreement on combinatorial units and formulas lessened its benefits. Hence, it could be argued that there was no improvement in the means to fix the reference of organic compound concepts, and even that there was a loss of precision. But starting in 1860, these shortcomings were slowly fixed. The synthesis of organic compounds accelerated, extending the landmark synthesis of urea of 1828 to acetic acid and a variety of sugars. These exemplary achievements managed to show that there was nothing intrinsically different between organic and inorganic compounds when it came to their capacity to be synthesized.[41] Also, it is important to stress that the inability to use the analysis–synthesis cycle involved a loss only relative to inorganic compounds. In comparison to the way in which the referent of animal and vegetable substances was fixed at the start of the century, even faulty formulas provided a better way.

Improvements in the combinatorial units used in formulas was a more complicated affair. Measuring the amounts of different alkalis that exactly neutralized a reference acid, or the amounts of different gaseous substances that combined with a reference gas, gave *indirect* information about the relative proportions of substances in the resulting compound: a neutral salt or an oxide. But to benefit from this, chemists had to make assumptions about the combining capacity of the reacting substances. If oxygen and hydrogen combined in one-to-one proportions, for example, then their proportions as ingredients would be the same as those in a compound like water. But if oxygen had the capacity to combine with twice the amount of hydrogen, then the two ratios would not coincide.

Assumptions about combining capacities, what later would be called "valency," were built into the calculations through the use of formulas: the formula HO assumed a one-to-one proportion, while H_2O assumed a two-to-one proportion. Thus, before the concept of valency was created, errors in the estimation of relative proportions were inevitable. When the density of gases was used to estimate combining proportions, another source of error was that many elementary substances exist in the gas state not as atoms but as molecules, and this conceptual distinction was not available to chemists until the second half of the century.[42]

Improving the accuracy of estimates of relative proportions was performed through a series of *mutual adjustments* between formulas and measurements, as well as between formulas, valency numbers, and the place of elementary substances in taxonomic schemas like the Periodic Table. Retrospectively, it is clear that the main value of atomic weights was not that they gave an accurate measure of the mass of single atoms but that they allowed *the ordering of elementary substances in a series* of increasing values. As the number of elementary substances multiplied, chemists noticed that these ordered series displayed periodic patterns: the chemical properties of the elements repeated at regular intervals, creating groups of substances bearing family resemblances to each other. These groupings were so remarkable that they could not be the product of chance. Moreover, the series displayed regular variations in valency across neighboring elements belonging to different groups. This implied that the numerical regularities in the Periodic Table and those in valency numbers could be used to correct the atomic weights assigned to various elements. In the end, it was this mutual adjustment guided by regularities in formulas, taxonomies, and direct measurements that allowed the units to slowly converge on a set of more accurate values.[43]

Formulas written with these improved units, in turn, helped to fix the referents of substance concepts in a better way. In addition, as we saw, the formulas themselves underwent changes that can be considered improvements. While empirical formulas were mere formalizations of quantitative statements about the chemical nature and proportions of the components of a substance, the other kinds progressively added more information: rational formulas singled out groups of components that played significant roles in reactions; type formulas captured recurrent motifs in the overall relations between components; while structural formulas added insight into the connectivity between components. This additional information could be used to solve referential difficulties, like those caused by substances that had identical composition but different properties. In addition, rational and type formulas acted as a guide to synthesis, leading to the development of series of substances in which something remained constant while something else varied. When these series were overlaid over the confusing jungle of organic compounds a significant improvement in their classification was achieved.

The concept of affinity enlarged by the concept of valency was also a clear enhancement over the concept of affinity alone. The extensions to the Affinity schema just discussed reflect the improvement. As we will see in the following chapter, before the century was over the Affinity schema would undergo other changes in the hands of physical chemists. Specifically, statement A-5" in the schema was changed from one about forces into one about the energy or work needed to perform a chemical transformation. The overall effect of these changes was that the concept of *selective attraction* was broken down into two components. On the one hand, the selective aspect of affinity became firmly linked to questions of spatial arrangement and mutual fit, while its power of attraction was reconceptualized in thermodynamic terms as depending on differences in chemical potential. Since each of these two aspects of the concept of affinity was handled by practitioners of different subfields, the schema itself reached a limit of improvability, and was not further developed. Or less metaphorically, the reasoning patterns used by organic and physical chemists could not be represented anymore using the schema.

We said in the Introduction that the evaluation of cognitive tools must be done both in terms of the improvements they have undergone in the past, as well as in terms of their *future improvability*. The advantage of ending a historical assessment in 1900, it was argued, is that we can use the track record of different cognitive tools over a century of further development as an added resource in our evaluations. What is the track record of the concept of valency, for example? Before answering this question we must emphasize that the world in which this concept was further developed was one in which some fundamental presuppositions of chemistry had been rendered problematic almost overnight. So if it can be shown that the concept of valency was able to survive and thrive in this new world, this should increase the cognitive value assigned to it retrospectively. The new world in question was one in which atoms had been shown not to be indivisible particles. In other words, substances that marked the limit of chemical analysis were shown to be capable of further *physical* analysis. The event that unleashed the avalanche of changes was the discovery of particles with a mass much smaller than that of the simplest chemical element, hydrogen. Electrons, as these *pieces of atom* were called, were confirmed to be charged particles in 1897, and almost immediately began to influence the way in which chemists thought about affinity, valency, and bonds.[44] The first quarter of the twentieth century is exemplary in this regard, because quantum mechanics did not yet exist and chemical practitioners contributed as much as physicists to the development of electronic models of the chemical bond.

The long experience with inorganic salts, the electrically neutral product of oppositely charged acids and bases, suggested that chemical bonds were electrostatic in nature, and the newly revealed existence of a negatively charged sub-atomic particle seemed to fit this *polar* model perfectly: a bond could be conceived as the end result of the transfer of electrons between substances, the substance that received the electron becoming more negative, the one donating it more positive. Such a polar model of the bond had been proposed as early as 1904.[45] On the other hand, chemists were aware that electrostatic attraction based on opposite polarities was bound to be relatively weak, and the bonds it created easy to break. But the bonds between carbon atoms forming chains and rings in organic compounds seemed too strong to be explainable in this way.

In addition, within the type approach to classification the universality of polarity had been successfully challenged. So in 1916 a different model was proposed, one involving not a transfer of electrons but the sharing of *pairs of electrons*. Because the sharing of negative charges leaves the overall electrical state unchanged, this other model was non-polar.[46] The colorful organo-metallic compounds displaying variable valencies were also modeled as involving non-polar bonds in which the shared pair was drawn from one and the same (metallic) atom. By 1927 it was clear that the two rival models of the bond, as well as the two rival versions of the Affinity schema, each captured part of the truth: some bonds were polar (or ionic) while others were non-polar (or covalent).[47] Retrospectively it can be argued that the variability of models and schemas turned out to reflect the variability of the objective phenomena.

What role did the concept of valency play in this evolution? As just discussed, valency numbers had been interacting with formulas and taxonomic schemas (the Periodic Table) in the last three decades of the nineteenth century, the numerical regularities in each of these cognitive tools serving to perform mutual adjustments and generate convergence. At the turn of the century, any model of atoms (and of the bonds these formed to make molecules) had to explain these numerical regularities. The number *eight*, in particular, appeared repeatedly in both the valency of compounds and in the periods of the table. For example, if the valency of the elements in stable compounds was added, the result tended to be a multiple of eight: the valency number for H_2O was four; the one for NH_3 was eight; for $MgCl_2$ was 16; and for $NaNO_3$ was 24. For highly-reactive compounds this regularity broke down, and this implied the existence of a connection between the number eight and the *stability* of a compound.[48] And similarly for the Periodic Table, its second and third rows, which originally contained seven elements, had been recently expanded to eight in response to new additions to the domain: the inert gases first isolated in 1894.[49] The name of these elementary substances reflected their weak disposition to enter into chemical combinations, and this implied another link between stability and the number eight.

The recurrence of the number eight suggested that the arrangement of electrons in an atom had the symmetry of a cube, so models in which the electrons of successive elements were arranged in concentric cubes were put forward as early as 1902.[50] These cubic models were discarded early on, but something survived: the idea that the ordering of the table could be used to guide *how to to fill successive layers* around a nucleus in a model of the atom.[51] This method of creating atomic models took time to be fully developed, but from the start it indicated that if electrons were indeed arranged concentrically, then the *outermost layer* had to be special: not only was this layer the part of one atom with which other atoms could interact, but its distance from the nucleus implied that its electrons should be less strongly bound and therefore more available to be shared or exchanged. In short, this outer layer, or shell, should be the locus for bonding phenomena. Between 1916 and 1919 this outer layer became the *valency shell*, while the number of electrons in the layer available for the formation of bonding pairs became an explanation for the combining power of elementary substances (valency number).[52]

Thus, it can be argued that the concept of valency was improvable, and that it guided the first steps into the world of invisible microscopic entities. And similarly for the concept of a chemical bond. On the other hand, there was improvement in these two cases only to the extent that the entities in terms of which the enhancement was made, electrons, turned out to be real. Track record is also an important consideration in this regard. In particular, unlike many entities dear to practitioners of sub-atomic physics, electrons have been around long enough, and the rays they form as they move (cathode rays) have been incorporated into so many devices, that we have a very good idea of their properties and capacities. Up to relatively recently, for example, all screens used in televisions and computer terminals were basically cathode ray tubes (CRTs). A CRT consists of an electron emitter and a fluorescent surface acting as a screen. Images are produced by deflecting the electron beam electrostatically or magnetically. There can be little argument that CRTs have greatly improved by their own standard: the quality of the image produced on the screen. But they have become better imaging devices thanks to our increased ability to causally manipulate electrons. And this, in turn, has been made possible by our greater understanding of the capacities of electrons to affect the phosphor-coated surfaces on which they are sprayed, as well as their capacity to be affected by electric or magnetic fields.[53]

Whether the concept denoted by the word "electron" has a referent or not must be settled by considerations such as these, and not by the possession of the "true theory" of electrons. In other words, the referent of the concept is fixed instrumentally, by our ability to produce beams of electrons and our ability to use these to successfully affect other things. Although cathode rays were at first only a laboratory phenomenon, an enigmatic glow within vacuum tubes, they were immediately converted into laboratory instruments used to probe the microstructure of other entities. Thus, statements about them, like "Electrons are particles," can be replaced by others like "Electrons are waves," while statements like "Electrons are the ultimate unit of electric charge" may be proved to be false, if quarks do turn out to have fractions of that charge. And yet our confidence that electrons exist can be sustained in the face of changes in our beliefs about them, if our ability to deploy them to reliably *produce other phenomena* improves, as it in fact has done.[54]

From Personal to Consensus Practice 1800–1900

1800–60

Jöns Jacob Berzelius (1779–1848), Joseph Louis Gay–Lussac (1778–1850). Justus Von Liebig (1803–73), Jean–Baptiste Dumas (1800–84),

Auguste Laurent (1808–53), Charles Gerhardt (1816–56), Alexander Williamson (1824–1904), William Odling (1829–1921), Edward Frankland (1825–99)

The study of plant and animal substances—resins, gums, waxes, syrups, oils, sugars, alcohols—was carried out in the eighteenth century alongside that of mineral substances, but the complexity of the organic portion of the domain did not allow the same rate of development and improvement as its inorganic counterpart. Lavoisier successfully broke down sugar, alcohol, wax, and olive oil into elementary substances producing the rather surprising result that they all consisted of carbon, oxygen, and hydrogen in different proportions. In the 1790s, Fourcroy added nitrogen to this small list of elementary components.[1] However, although these constituted real achievements in organic analysis, the useful properties of organic substances disappeared once the level of elementary substances was reached. As we noted in the previous chapter, the concept of a compositional hierarchy helped to solve this dilemma by supplying a justification for why analysis should cease at a given point. Not elements, but more *proximate principles*, that is, substances higher in the hierarchy, were the core of the organic domain at the start of the nineteenth century.

On the other hand, the fact that the list of basic components of substances of plant and animal origin was so small did present chemists with a problem: why does such a small set of elements produce such a large variety of compounds? One early proposal was that, unlike the binary compounds exemplified by neutral salts, substances of plant or animal origin were ternary or even quaternary, and that the affinities involved were also multiple. This implied that the balance of forces used to explain the outcome of affinity-mediated reactions had to be more precarious than that of mineral substances, explaining why analysis by fire led to the loss of the medicinal virtues of organic substances: the equilibrium between three or four affinities could only be achieved at low temperatures and it was destroyed by intense fire.[2] To this daunting complexity it must be added that, at the beginning of the century, organic compounds could be analyzed but not resynthesized. This not only left the results of analysis unconfirmed, but it suggested that only biological organisms could act as the "laboratories" in which such a synthesis could be carried out.[3] The uncertainty created by this situation meant that for the first two decades of the nineteenth century organic substances were classified using traditional properties and dispositions: their animal or vegetable source, their sensible qualities, and their chemical behavior.

One of the main challenges faced by chemists was therefore to develop new methods of classification to bring this part of the domain into harmony with the part represented by the neutral salts. Much of the variability in personal practices in this period was caused by different approaches to classification. But there were other sources. First, there were the different attitudes of chemists towards *vitalism*: should they classify artificially-produced carbon-based substances together with those extracted directly from plant and animals? Some practitioners (Berzelius, Liebig) refused to group them together, while others (Dumas, Laurent, Gerhardt) had no problem with that.[4] Variation in personal practices was also related to the combinatorial units used in formulas (equivalents, atomic weights), as well as by the conventional values used for the reference substances: Berzelius used carbon = 12 and oxygen = 16; Liebig used carbon = 6 and oxygen = 8; Dumas used carbon = 6 and oxygen = 16.[5] Finally, there was variation due to differing ontological commitments: some were atomists (Laurent, Berzelius, Williamson) while others refused to accept invisible particles (Dumas, Gerhardt, Odling).

The personal practice of Jöns Jacob Berzelius was informed by two related concerns: to find an electrochemical explanation of affinity and to devise a classification schema using his chemical formulas. In his own experiments with the Volta pile in 1804, Berzelius had observed that when a continuous current was passed through a dissolved compound, the compound was not only decomposed, but the products of its dissociation accumulated in the negative and positive poles. This phenomenon could be explained by thinking of the products of the reaction as themselves electrically charged. Oxygen, for example, was strongly attracted to the positive electrode so it could be considered to be electronegative, while potassium was drawn equally strongly to the negative side, and could be considered electropositive. Taking the behavior of these two substances to represent the ends of a continuum, other elementary substances could be placed in between: sulphur, for example, behaved like a positive element in interaction with oxygen, but as a negative one in interaction with metals, so it could be placed at an intermediate position.[6] These results strongly suggested that the forces of affinity were electrical in nature, and that their intensity accounted for the different degrees of attraction exhibited by different substances. Moreover, the known binary composition of neutral salts (as well as that of acids and alkalis) could now be explained as the coupling of positively and negatively charged ingredients. Finally, this explanation could be extended *by analogy* to organic substances, eliminating the idea that their affinities were ternary or quaternary.

Once Berzelius had developed his formulas in 1813 the analogy could be made more precise. Thus, an inorganic substance like sulphuric acid,

H_2SO_4, could be modeled as a binary compound if we rearranged its symbols like this: $SO_3 + H_2O$. In a similar way an organic substance like acetic acid could be modeled as $(C_4H_6) O_3 + H_2O$.[7] This allowed Berzelius to think of the group C_4H_6 as playing a role analogous to that of sulphur in sulphuric acid, that is, the role of a basic building block. If a chemical reaction was conceived as an interaction in which elementary substances remained unchanged while compounds were transformed, the fact that the group C_4H_6 remained unchanged made it a kind of "organic element," or as these relatively constant groups were called, a *radical*.

This analogy provided a bridge between the part of the domain that had been thoroughly mastered (neutral salts) to that which presented chemists with the greatest challenges. The new formulas supplied a way to make this bridge more tangible, but they remained part of Berzelius' personal practice well into the 1820s. French chemists were the first to adopt them, while German chemists waited until the 1830s to begin using them.

The *integer* numbers that occurred in the new formulas could be justified by a series of results, one of which was made by a contemporary of Berzelius: Joseph Louis Gay-Lussac. Prior to the work of Gay-Lussac it had already been established that when two elementary substances combine they do so either in constant proportions or in integral multiples of those proportions. This empirical generalization was derived from quantitative studies that focused on the weight of substances, that is, gravimetric studies. Gay-Lussac's approach was different. He was a pioneer in the use of *volume* measurements in chemical analysis, so his personal practice became oriented towards the *titrimetry*, the volumetric study of substances in the gaseous state. As part of this line of research he discovered in 1808 that when two gases combined they tended to do so in integral proportions: 100 volumes of carbon monoxide combined with 50 volumes of oxygen yielded 100 volumes of carbon monoxide; and similarly, 100 volumes of nitrogen combined with 300 volumes of hydrogen formed 200 volumes of ammonia.[8] This striking regularity provided further evidence that the numerical relations between elements in a compound exhibited *discontinuities*, making it reasonable to expect that intermediate values did not exist. As it was argued in the previous section, the elimination of these intermediate substances—substances without a well-defined stoichiometric identity—was an important step in the consolidation of the boundaries of the organic domain.[9]

Gay-Lussac's volumetric approach, more accurate that distillation as a form of analysis, also allowed him to contribute to another landmark result, starting in 1815. In the first half of the nineteenth century, alcohol and ether had become *model substances*, playing a role similar to that of oxygen in the last decades of the 1700s. Early analyses of alcohol and ether had shown that they were both composed of water and a gaseous substance (ethylene). It was also known that ether could be produced by reacting alcohol with sulphuric acid. Quantitative analysis showed that the only difference between the two was in the amount of water that entered into their composition, and this led to modeling that chemical reaction as one in which sulphuric acid removed water from alcohol to yield ether as a product. Gay-Lussac confirmed that *dehydration* was a plausible mechanism, and that the volumes of the substances in the reaction exhibited the requisite discontinuities.[10] This had important consequences because if water and ethylene were in fact components of alcohol and ether, this implied that substances extracted from organic materials could be made of inorganic building blocks. Those who, like Berzelius, tended towards some form of vitalism could not accept this result. But to others this conclusion held the promise that the organic and the inorganic parts of the domain could one day be unified.[11]

Organic analysis, as practiced by chemists like Berzelius and Gay-Lussac, consisted in the production through combustion of two main products: carbon dioxide and water. From these two, the quantities of carbon and hydrogen in the sample being analyzed could be directly determined, while the quantity of oxygen could be indirectly calculated. Using this approach, Gay-Lussac (together with Louis Jacques Thenard) had successfully analyzed 19 different organic substances by 1811, while Berzelius had analyzed another 13 by 1814.[12] Both of these chemists improved on existing techniques in what amounted to an unplanned collaboration: Berzelius improved the apparatus while Gay-Lussac discovered better oxidants to carry on combustion.[13] However, the amount of plant or animal substances that they could analyze was small because the pneumatic trough they used could not collect the large amounts of gas produced. This limited the accuracy of the results. Another limitation was that the nitrogen content of the sample had to be determined separately. In the early 1830s, Justus von Liebig, a former student of Gay-Lussac, created a new instrument that would surmount these difficulties, as well as reducing the considerable amount of skill needed for the performance of the analyses. Instead of gathering carbon dioxide as a gas, Liebig devised a way of capturing it in the condensed state in a series of glass-blown bulbs. This allowed the sample size, and the precision of the results, to be greatly increased.[14]

The *Kaliapparat*, as the new instrument was known, was not Liebig's only contribution to organic chemistry. As part of his personal practice he made important improvements to the radical approach to classification. In

1838, he gave a formal definition of the concept: a group of atoms is a radical if it is a constant component of a series of compounds; if it is replaceable in those compounds by a simple element; and if the elements with which it forms compounds are replaceable by equivalents of other elements.[15] He was also the first to point out that which groups stay constant varies depending on the chemical reaction chosen as standard, so that there was an element of conventionality in the determination of radicals.[16] Hence the analogy between elements and radicals was not perfect, since elementary substances do not change status depending on context. The analogy also broke down in a different way: in Liebig's time most radicals could not be separated and purified, preventing the fixing of the referent of terms like "etherin" or "benzoil" instrumentally, a function performed by the more tenuous and indirect means of formulas. This was only a temporary difficulty—later chemists were able to isolate some radicals and explain why those that resisted isolation did so—but Liebig had improved the analogy through his careful qualifications.

Liebig joined forces with his rival Jean-Baptiste Dumas to publish a manifesto on the radical approach in which the authors emphasized the role played by radicals in allowing organic substances to be modeled as binary compounds. The alliance, however, did not last long because new phenomena began to create problems for the electrochemical account of binary composition. The first signs that something was wrong came not from the phenomenon itself—a chemical reaction in which two novel substances, chloral and chloroform, were synthesized—but from a new model of the reaction based on Berzelius' formulas created by Dumas.[17] The model indicated that an electropositive element (hydrogen) had been replaced by an electronegative one (chlorine), a replacement that should have been impossible according to the radical approach. Dumas was able to show that hydrogen was replaceable not only by chlorine, but that in other reactions it was similarly replaceable by bromine and iodine which were also negatively charged.[18] This discovery displayed the power of formulas to generate insights into chemical reactions: the manipulation of symbols on paper appeared to be an effective means to understand the possible recombinations that a substance's components could undergo in actual laboratory reactions. If this turned out to be correct, models using formulas could compensate for the fact that organic reactions tended to exhibit a cascade of transformations in which a variety of byproducts were formed. These byproducts were so hard to isolate, and their masses so hard to measure, that the old balance-sheet approach to quantitative analysis was too difficult to apply. Better to balance combinatorial units than messy laboratory quantities.[19]

Dumas's young assistant, Auguste Laurent, followed at first in his master's footsteps, studying complex organic reactions by manipulating formulas. He explored the different substances that could be generated from naphthalene by replacing hydrogens with chlorine, bromine, and oxygen, making sure naphthalene's 28 units remained constant in the formula models.[20] To generate this series of substances he used *substitution reactions*, rather than the displacement reactions common in the radical approach.[21] Soon after that, Laurent began to search for an alternative to that approach, using a cognitive resource that was unique to his personal practice: crystallography. From his background in this field he knew of an unexplained phenomenon in which the salts formed by different metals had the tendency to crystallize in similar shapes. The phenomenon of *isomorphism*, as it was referred to, suggested that the shape of crystals was to a certain extent independent of the nature of the metallic components of a salt, and was related instead to the proportions and groupings of these components as a whole.[22] This insight developed into a *geometrical* approach to compositional problems when Laurent discovered that one of the compounds he had obtained from naphthalene displayed isomorphism. In 1837, Laurent created the first spatial model of substitution, by imagining that a basic hydrocarbon (C_8H_{12}) was a prism with eight corners occupied by carbon atoms and 12 edges occupied by hydrogen atoms. Adding extra hydrogen or chlorine atoms generated a pyramidal form (not isomorphous), but replacing a hydrogen atom by a chlorine atom maintained the prismatic form, and hence, yielded an isomorphic crystal shape.[23]

Laurent had no problem talking of atoms and molecules because like Berzelius (and unlike many other chemists) he was an atomist. In 1846 he gave the first unambiguous definition of these two terms: an atom is the smallest quantity that can exist in combination, while a molecule is the smallest quantity that can bring about a combination.[24] Although neither his geometric hypothesis nor his distinction between atoms and molecules had any immediate impact on the community, they did affect his personal practice, focusing it on the entire molecule, not groups of atoms, as the key to taxonomy. A new classification schema could be created, he thought, if differently composed substances can nevertheless belong to the same *type*. This conclusion would be greatly strengthened if substances were found that possessed not only similar physical properties (crystal shape) but also similar chemical properties. Through manipulation of formulas, Laurent predicted that acetic acid ($C_4H_6O_3$) + H_2O could be transformed into trichloracetic acid through a substitution of six hydrogen atoms by the same number of chlorine ones, and that the two substances would possess similar chemical behavior. In 1838, Dumas carried out the necessary experiment, a simple reaction of acetic acid with chlorine that could be easily reproduced, and verified the prediction. This consolidated the type approach to classification: two substances could be said to belong to the same type if they had the same number of combinatorial units (equivalents) and similar physical and chemical properties.[25] Unfortunately for Laurent, Dumas was a much more prestigious chemist, working in a fully equipped laboratory in Paris, instead of in the provinces without any resources, so he won the priority dispute that ensued.

But Laurent found an ally in Charles Gerhardt, who complemented his ability to predict the existence of substances using formula models with an equally impressive skill for synthesizing them.[26] Gerhardt was an avowed anti-atomist but this did not prevent an enduring and fruitful collaboration, illustrating once more that opposite ontologies do not lead to incommensurable world-views: Laurent could conceive of substitution reactions as literally involving the replacement of microscopic entities, while Gerhardt was free to conceptualize them as revealing relationships between macroscopic entities that were useful for classification.[27] On the other hand, Gerhardt realized that existing classifications of organic substances depended crucially on the choice of combinatorial units, and that the conventions used to generate them were proliferating out of control. Atomic weights, for example, could be determined using as a standard the two volumes occupied by two grams of hydrogen, or an alternative four-volume method which doubled the number on units.

Thus, water's formula using the first convention was H_2O, while using the second yielded H_4O_2. If equivalent weights were used then the formula became HO.[28] As with any other conventional unit of measure, this was not in itself a problem as long as chemists had the means to translate one set of formulas into another, like translating degrees of temperature from centigrade to fahrenheit units. Nevertheless, there were good reasons to try to get the community to converge on a single standard. Traditionally, the formulas for inorganic substances were written using two-volume atomic weights, while those for most organic substances used a four-volume standard. Thus, one reason to advocate a reform of conventions was to harmonize units across the entire chemical domain, something that could be achieved by simply doubling the atomic weight of carbon and oxygen— carbon was assigned the number 12 instead of 6, and oxygen the number 16 instead of 8—and by reducing by half the atomic weight of certain metals. These suggestions allowed Gerhardt to play a central role in the reform of combinatorial units in the 1850s.[29]

The value of the conventions that emerged from this reform can be illustrated by the role they played in the solution of an old problem. As mentioned above, alcohol and ether had become model substances, and the chemical reaction in which ether was produced by the dehydration of alcohol had become an exemplary case of what can be achieved through the use of formula models. But the model had a problem: the formula for alcohol, like that of other organic substances, was written using a four-volume standard, while the formula for ether used the alternative two-volume standard. This was a glaring inconsistency, but it was necessary to make the model work. Laurent and Gerhardt realized that a better model for the reaction could be created by using two-volume formulas for both substances, reducing the number of units of alcohol by half, but compensating for this by assuming that *two alcohol units* were involved in the production of one unit of ether.[30] Evidence for the validity of this alternative model was not easy to produce because the reaction that used sulphuric acid as dehydration agent could not differentiate between the two rivals. The British chemist Alexander Williamson resolved the stand-off by designing a reaction that could: he replaced the original reactants with a set of closely related ones that also produced ether, but such that their formula models predicted different outcomes—the traditional model predicted the production of two molecules of ordinary ether (or diethyl ether), while the new model predicted a single asymmetric form of ether (ethyl-methyl ether). When Williamson carried out the reaction in his laboratory he confirmed the latter's prediction.[31]

The contributions of Laurent, Gerhardt, and Williamson resulted in a new kind of chemical formula (type formulas) and a new taxonomic schema to rival the radical approach. As we discussed in the previous section of this chapter, type formulas attempted to capture similarities in the behavior of different substances under substitution reactions. Thus, substances that underwent similar substitutional changes as water were classified as belonging to *the water type*, the formula for which was created by Williamson following an initial proposal from Laurent.[32] The formulas were explicitly designed to display these family resemblances: alcohol and ether, for example, were classified as belonging to the water type using formulas like these:

$$\left. \begin{matrix} H \\ H \end{matrix} \right\} O \qquad \left. \begin{matrix} C_2H_5 \\ C_2H_5 \end{matrix} \right\} O \qquad \left. \begin{matrix} C_2H_5 \\ H \end{matrix} \right\} O$$

Similarly, the *ammonia type,* based on the empirical formula for ammonia, H_3N, could be used to exhibit the kinship of ammonia with substances like ethylamine and methylamine like this:

$$\left. \begin{array}{l} H \\ H \\ H \end{array} \right\} N \qquad \left. \begin{array}{l} C_2H_5 \\ H \\ H \end{array} \right\} N \qquad \left. \begin{array}{l} CH_3 \\ H \\ H \end{array} \right\} N$$

What exactly these new formulas were assumed to be varied with the ontological commitments of a chemist: the formulas could be seen as capturing information about the spatial arrangement of elements in a molecule, or as constituting mere records of the history of substitutions that a substance could undergo. Williamson had begun his career as an anti-atomist, but he eventually changed his position and had no problem viewing the oxygen atom to the right of the curly bracket in the water type formulas as literally *holding together* the elements or radicals on the left side. To him the type formulas were related to the structure of invisible molecules like an orrery is to a planetary system.[33] This stance towards type formulas remained part of Williamson's personal practice for a long time. But another one of his contributions generated a line of research that would eventually encompass the entire community. Some type formulas, like that for sulphuric acid, showed that of all the various substitutions that substances could undergo, the substitution of single hydrogens was special. In particular, elementary substances, as well as radicals, could be classified by the number of hydrogen atoms they could replace.[34] Thinking about *hydrogen substitutability* as a characteristic disposition of chemical substances was an early formulation of the idea that elements or radicals have a capacity to combine *with a well-defined number* of other elements or radicals.

The name for the concept referring to this disposition remained in flux for decades, the terms "basicity," "atomicity," "quantivalence," and "valency" used by different authors.[35] But the importance of the concept itself was quickly grasped by chemists like William Odling. Odling not only recognized the universality of these quantitative dispositions, but he introduced a way to represent them in type formulas. The symbol for a substance like potassium, for example, was written as K' to indicate that it could substitute a single atom of hydrogen, while the ones for tin and bismuth were rendered as Sn" and Bi'" to denote their capacity to replace two and three hydrogen atoms respectively. Like Gerhardt, Odling was an anti-atomist, so for him the number of prime symbols associated with an element or radical was just that, a number. This number, to be sure, captured an important regularity that was useful for classification, but it did not carry any commitments to explain it either in macroscopic terms, using affinity relations, nor in microscopic terms, using bonds.[36] In 1855, to classify substances like methane, Odling introduced a new type, *the marsh gas type*, the formula for which had long-lasting consequences for organic chemistry because it suggested that carbon had the capacity to replace four hydrogens, that is, that carbon was *tetravalent*.[37]

The full development of the marsh gas type, and of the multiple consequences of the tetravalency of carbon, was the achievement of another chemist, August Kekulé, whose work will be discussed below. At this point it is important to emphasize that in the development of the concept of valency not only the advocates of the type approach were involved, but also those who continued to defend the radical approach, such as the English chemist Edward Frankland. It was acknowledged by both sides that the existence of persistent groups was an important fact, but only in the radical approach was it imperative that these groups be isolatable in the laboratory. Frankland unsuccessfully worked on the isolation of the alkyl and ethyl radicals, but in the process he obtained a byproduct, zinc ethyl, that was the first organo-metallic substance to join the domain. As he studied these hybrids he noticed that their metallic components had a maximum combining capacity.[38] In 1852 he extended this idea to non-metallic elements and added the important observation that this combining capacity displayed numerical regularities: nitrogen, phosphorous, antimony, and arsenic, for example, formed compounds that contained three of five combinatorial units of other elements, as if their affinities were optimally satisfied in just those proportions.[39] He coined the term "atomicity" for this combining power, a term that would eventually become synonymous with "basicity," the name used in the type approach for the capacity of elements or radicals to substitute for hydrogens. Thus, each of the rival approaches to classification had something to contribute to the development of valency.

Let's examine now how these diverse personal practices converged on a consensus practice. The formation of a new consensus is normally dated to 1860, the year in which the first international conference on chemistry was held at Karlsruhe, Germany. But as it has recently been shown, the relative ease with which a consensus was forged at the event was the result of a quiet accumulation of changes that had taken place in the previous decade, among which was Williamson's synthesis of ether in 1850; Gerhardt's codification of type theory in 1853; Odling's treatment of the marsh gas type in 1855; and Frankland's creation of the concept of atomicity in 1852. This slow preparation was necessary because none of the older chemists with the authority to block change, such as Liebig and Dumas, accepted the reforms. This rejection implied that siding with the reformers (Laurent and Gerhardt) went against the professional interests of young practitioners, who endangered their careers by accepting the ideas of two chemists who had been ostracized until very recently.[40] As in the controversy over phlogiston, the consensus that emerged was a partial one. No agreement was reached on ontological matters, the strong atomist positions of Laurent and Williamson taking another 50 years to become widespread.[41] And as in that earlier controversy, advocates of the radical approach who retrospectively seem to have held on to their views longer than it was reasonable, played a constructive role as critics, forcing rivals to sharpen their positions and preventing useful variations from being eliminated prematurely.[42]

We can get a sense of the new consensus by examining a textbook published a few years after the Karlsruhe conference.[43] If we compare this textbook to the one used to sample the state of consensus practice in 1800, the most striking difference is the approach used to classify and identify organic substances. At the start of the century, gums, sugars, oils, gelatins, as well as blood, milk, saliva and other bodily fluids, were classified on the basis of their plant and animal sources.[44] The bodies of living creatures were considered the laboratories in which the difficult synthesis of organic compounds was achieved, so these are discussed in great anatomical and physiological detail. By 1860, the idea that organic chemistry is the study of carbon compounds, *regardless of their source*, is well established. Of these, compounds of carbon and hydrogen are considered at this point the key to the mysteries of the organic domain.[45]

The ordering of this realm is performed using a *serial taxonomy* that disregards any of the criteria used in 1800, one that focuses exclusively on the chemical reactions that produce the series of kindred substances. Of these, the series comprising the substances we know as methane, ethane, propane, butane, pentane, and so on, has been particularly well worked out. By rendering this series with formulas, its internal structure can be clearly displayed: CH_4, C_2H_6, C_3H_8, C_4H_{10}, C_5H_{12}, C_6H_{14}, C_7H_{16}, C_8H_{18}, C_9H_{20}. Even a cursory examination of this series shows that it increases by a modular amount, CH_2, and that its structure is captured by the formula C_nH_{2n+2}.[46]

A variety of cognitive tools had to be improved (or invented) to make this dramatic transition possible. First of all, there is the cluster of concepts and statements associated with the practice of stoichiometry. If carbon had become the center of the organic domain, its frontier was delineated by compounds that displayed sharp discontinuities in the proportions of their components. Some statements, like statement 1, were already included in the previous textbook, but when that teaching aid was published the universality of discontinuities was still debated.[47] In 1860 not only has statement 1 been fully accepted, but it has been joined by others that provided further evidence for its truth:

1 Two substances combine in either one determinate proportion, or in several proportions that are integer multiples of one another.[48]

2 The volumes of two elementary substances in the gas state always combine in integral proportions.[49]

The determination of combinatorial units, equivalents, and atomic weights, although still plagued by uncertainties, has been improved by the discovery of correlations between the units and other measurable properties (specific heat, crystal shape) and by improvements to the concept of the gas state. The correlations are expressed by statements like these:

3 Substances with isomorphous crystal shapes have compositions expressible by analogous formulas.[50]

4 Multiplying the specific heat and the atomic weight of a given elementary substance yields a constant number.[51]

These correlations provided constraints on the choice of value for combinatorial units. As was suggested at the end of the previous section of this chapter, units, formulas, classifications, and a variety of measured properties, were mutually adjusted to one another, slowly converging on more accurate values.[52] But in addition to reciprocal corrections, the improvement of combinatorial units involved a major conceptual leap regarding gaseous substances. In particular, elementary substances like sulphur, phosphorous, or arsenic yielded values two or three times greater when their atomic weights were calculated using statement 2, as opposed to the values obtained using statement 4.[53] This discrepancy could be corrected if a gas was conceived not as being an elementary substance, but *a compound of the same element*. Pure oxygen and nitrogen gases, for example, had to be thought of as a compound of two units of these elements, and pure phosphorous and arsenic gases as compounds of four units. Expressed in atomistic language, the conceptual change required thinking of pure elementary gases as composed not of single atoms, but of diatomic or tetratomic *molecules*.[54] This idea not only repelled anti-atomists, but clashed with the electrochemical model of affinity, since it implied that elements with the same electrical charge could enter into a stable union. Despite the resistance, the improved concept of gases has entered the 1860 consensus, as has a powerful statement neglected for almost 50 years:

5 At the same temperature and pressure, a given volume of any gaseous substance contains the same number of molecules.[55]

In addition to this cluster of concepts and statements providing the units and formulas for taxonomic series like CH_4, C_2H_6, C_3H_8, and so on, there was another cluster supplying an explanation for the series, as well as providing a solution to the most pressing problem of organic chemistry: Given that organic compounds are all made of three or four basic elements, why is there such an enormous variety of compounds?

This was the cluster that formed around the concept of *valency*. In 1800 the problem of the variety of organic compounds was approached by invoking the existence of a delicate balance of forces in the bodies of plants and animals, a balance of ternary or quaternary affinities. In 1860, the explanation is still given in terms of affinity, but the latter has ceased to be a unitary force and has been *quantized into units*. Elements were thought to possess a different number of *affinity units*, or as anti-atomists preferred to phrase it, a different valency number. Either way, the key to the solution gravitated around capacities to combine. With the exception of hydrogen, the building blocks of organic substances were known to be capable of combining with more than one element, that is, they were all *polyvalent*.[56] Thus the new solution to the original problem was framed using this concept, alongside statements like these:

6 Elementary hydrogen is monovalent.

7 Elementary oxygen is divalent.

8 Elementary carbon is tetravalent.

The tetravalency of carbon was special because carbon could use its affinity units to combine not only with four monovalent elements or radicals, but also *to combine with itself*. The first member of the methane series, CH_4, illustrates the former capacity, carbon using its four affinity units to bind four hydrogens, while the second member, C_2H_6, illustrates the latter: the two carbons have eight affinity units, six used to combine with the hydrogens while the other two are used by the carbons to join with each other.[57] Those compounds in which all affinity units are used are said to be in a *state of saturation*, a new concept but one related to the eighteenth-century notion of a saturation point, the point at which the affinities of an acid and an alkali are best satisfied, producing a neutral salt. By analogy, a compound in the saturated state was defined as one in which all the affinity units have been satisfied. This condition makes a saturated compound incapable of entering into combination with other elements by addition or subtraction. The only way such a compound can change is through substitution reactions.[58] The concept of saturation is also used to solve a problem concerning the other building blocks of organic substances: radicals.[59] The problem can be expressed like this: Given that radicals, like elements, have an independent existence, why are they so hard to isolate and purify?

The most significant factor in the proposed solution is the number of unsatisfied affinity units in the radical. Unsaturated radicals that have an *even number* of unsatisfied units, such as ethylene (C_2H_4)", can exist in a free state and combine with elements like oxygen in the same way that metals do. Unsaturated radicals possessing an *odd number* of units that are unsatisfied, such as ethyl (C_2H_5)', cannot exist in a free state, and this explains why they cannot be isolated, with one exception: ethyl can be isolated if it forms a diatomic molecule with itself, yielding butane or C_4H_{10}.[60]

The 1860 consensus contained a variety of less pressing problems posed by known exceptions to newly accepted general statements. Thus, even after greatly refining the determination of specific heats to rule out some anomalous cases, statement 4 above continued to fail for carbon, silicon, and boron. Carbon, in particular, refused to comply with the general pattern, posing the following problems:

> Given that charcoal, graphite, and diamond have the same atomic weight, why do they possess different specific heats?
> Given that carbon does not change atomic weight in a compound substance, why does it possess a different specific heat as an elementary substance and as part of compounds?[61]

The new consensus also contained changes in explanatory devices. The textbook acknowledges the dual role played by formulas: to classify substances and to model chemical reactions.[62] A rational formula or a type formula is a record of the behavior of substances during reactions, registering the tendency of a compound to break in certain directions but not others. Thus, when a formula is used to model a chemical reaction, the information it contains about preferential ways of breaking can be used to generate hypothetical mechanisms about the transformation.[63] Formulas themselves exhibited regularities, and these also demanded explanation. Type formulas, for example, revealed that a large number of organic compounds could be classified under a small number of types, leading defenders of the rival approach to ask why this should be so. The textbook approaches the answer using the combinatorial capacities of the elements on the right side of the brackets: the existence of the water type can be explained by oxygen's possession of two affinity units, while that of the ammonia type is accounted for by nitrogen's possession of three affinity units.[64] Formulating explanations along these lines has affected the reasoning styles of atomist chemists, an effect that can be captured by making modifications to the Affinity schema, adding a qualification to its first statement:

A-1. Substances possess dispositions to selectively combine with other substances.
A-1a. Combinations are formed when the affinity units of one element or radical are satisfied by those of another element or radical.[65]

In the last four decades of the nineteenth century, statement A-1 acquired even more qualifications, as new concepts began to emerge, like the concept of a chemical bond involving the mutual satisfaction of two affinity units, as well as the idea that affinity forces have a direction in space. These modifications became a consensus only among atomists, but the influence of the latter increased steadily until they became the majority around 1911.[66]

1860–1900

August Kekulé (1829–96), Johannes Wislicenus (1835–1902), Archibald Scott Couper (1831–92), Alexander Crum Brown (1838–1922), Jacobus van't Hoff (1852–1911), Adolf von Baeyer (1835–1917), Marcellin Berthelot (1827–1907), Emile Fischer (1852–1909)

Much like alcohol and ether were the exemplary substances of the first half of the nineteenth century, *lactic acid and benzene* became the model substances of the second half. Neither substance was new and their empirical formulas were long known, but both continued to pose problems by their relationship to *isomers*: substances having the same composition but different properties. The investigation of the isomers of benzene (and those of its derivatives) had revealed numerical patterns that had become firmly established in the 1860s. From experiments with substitution reactions it was known that if only one hydrogen in benzene was replaced by a radical, a single isomeric variant was obtained. If two hydrogens were replaced, then three different isomers were produced. Using formulas, and starting with benzene's empirical formula, C_6H_6, these findings could be expressed like this: replacing one hydrogen yielded a single isomeric variant of C_6H_5X (no matter what monovalent radical X was), while two hydrogen replacements yielded three variants of C_6H_4XY, regardless of what X and Y were.[67] The multiplicity of substances that could be derived from benzene, and the regular patterns its isomers formed, ensured its prominence as the parent substance of the increasingly important chemical family known as *aromatics*.

Lactic acid's role as model substance was also related to questions of isomerism. This substance had been part of the domain since 1780, deriving its name from the fact that it was produced from sour milk. A variant from animal muscle, isolated in the nineteenth century, displayed *optical isomerism*: its composition was identical to the older substance but it interacted with light in a different way. Optical isomerism had already been observed in other substances. One isomer of crystallized tartaric acid, for example, could rotate the plane of vibration of light in one direction, while another rotated it in the opposite direction. Although this remarkable phenomenon had been discovered in the 1840s, there were not enough instances of it to make its explanation urgent. Further reducing the incentives to account for it was the fact that optical activity (like crystal shape or boiling point) was used only as a means to identify substances, not as a chemical property that demanded explanation.[68] Nevertheless, in the 1860s interest in the optical behavior of the isomers of lactic acid increased, leading to the evolution of *spatial* models in an attempt to explain this behavior, or at least to predict its occurrence.

Let's begin the discussion of the personal practices of this period with the work of the German chemist August Kekulé. Prior to the period under discussion, Kekulé had already introduced several ideas that were included in the consensus of 1860: that the valency number implied the quantization of affinity forces; that what he referred to as "affinity units" could be used to think about chemical bonds; and that carbon could use some of its units to bind with itself. This had led him to introduce the first instance of a spatial structure into chemistry: the *linear chains* that carbons could form as they repeatedly bonded with each other.[69] After 1860 he applied a similar line of thought to decipher the structure of benzene. From its empirical formula, H_6C_6, Kekulé could infer that its six carbons had 24 affinity units. But if six units were used to bind the hydrogens, and ten units to create the chain of carbons, what happened to the remaining eight? He hypothesized that the chain was closed, the two carbons at either end using two units to form a *hexagonal ring*. The remaining units were accounted for by a further hypothesis: the existence of carbon-to-carbon double bonds.[70] This was the origin of the now famous, and still enormously fruitful, hexagonal formula for benzene displaying alternating double bonds:

Kekulé defended his two hypotheses by pointing to the ability of his model to explain the relation between the number of substitutions and number of isomers. For example, the symmetry of the hexagon implied that, in the case of double substitutions, there were only three possible arrangements: a replacement of two neighboring hydrogens; of two hydrogens separated by one atom; and of two hydrogens separated by two atoms.[71] Reasoning along these lines, he could account for the fact that performing two substitutions on benzene always produced three isomers.

Another German chemist, Johannes Wislicenus, conducted the first in-depth analysis of the isomers of lactic acid. He began by exploring a problem that this substance had already posed to previous practitioners: in some chemical reactions it behaved like an alcohol, yielding ether as a product, while in other reactions it behaved like an acid, combining with bases to form acid salts. Neither the radical nor the type approaches to classification could explain this dual behavior. Wislicenus introduced a mixed type category to derive a new formula for lactic acid, a formula that suggested a different route for its synthesis. When the procedure was carried out it revealed that the problematic behavior was due to the existence of two isomeric variants.[72] Using Gerhardt's type formulas, Wislicenus could not display graphically the difference between the two variants, but given that they had an identical composition he speculated that the difference between alcoholic and acidic behavior might be caused by the relative position of a common radical. When structural formulas were introduced a few years later, a proper graphic representation of his hypothesis became available. Wislicenus went on to identify two more isomers of lactic acid, one of which displayed the enigmatic rotatory optical behavior.[73]

Structural formulas were created by a couple of Scottish chemists, Archibald Scott Couper and Alexander Crum Brown.[74] Odling, as we saw, modified type formulas to represent the number of affinity units, but Couper went beyond that and explicitly represented pairs of mutually saturated affinity units, that is, *chemical bonds*. He used dotted lines to do this. His structural formulas were capable of expressing graphically the concepts that carbon was tetravalent and that it had the capacity to bind with itself to yield chains. In addition, his formulas could be used to rule out the existence of isomers that type formulas suggested were real possibilities.[75] The structural formulas that became the preferred way to model isomers were the ones created by Crum Brown in 1864. He used Berzelian symbols enclosed within circles to represent atoms, and solid lines to represent bonds, but he explicitly stated that his formulas were not iconic, that is, they did not try to capture the spatial distribution of physical atoms in space, but were only symbolic expressions of the *connectivity* of the components of a substance according to the rules of valency. Soon after their introduction, his formulas became the basis for sculptural models in which croquet balls painted with different colors stood for elements, while metallic tubes played the role of bonds.[76] These sculptures are the ancestors of the ball-and-stick models still in use today, but like their flat versions they were still *symbolic*: in both cases the formulas stood for their referents by means of conventions.

It was as part of the personal practice of the Dutch chemist Jacobus van't Hoff that structural formulas finally became truly *iconic*.[77] Others had played with the idea that the four affinity units of carbon could be represented three-dimensionally, and since a triangular pyramid is the simplest polyhedron that contains four vertices, others before him had already proposed a *tetrahedron* as a possible spatial model. But van't Hoff was the first one to elaborate this hypothesis in order to solve a real problem, the problem posed by the isomers of lactic acid. He worked exclusively with previously published results, performing no syntheses of his own, but he was able to detect an important regularity in the mass of data: all known substances that had optically active isomers contained a carbon atom that was bonded to *four different* elements or radicals. In the case of lactic acid, $C_3H_6O_3$, this condition could be represented by the following rational formula:

$$C (H) (OH) (CO_2H) (CH_3)$$

This rational formula expressed the idea that the four affinity units of the first carbon were satisfied by bonds with one hydrogen atom and with three different radicals: a hydroxyl, a carboxyl, and a methyl. Van't Hoff improved on this formula by placing the carbon at the center of a tetrahedron, its affinities modeled as if they were directed towards its four corners, and then placing the hydrogen and the three radicals at the vertices. Once this was done, he was able see that the tetrahedron and its mirror-image could not be superimposed on each other, that is, that a triangular pyramid with differentiated vertices did not remain invariant under a mirror-image transformation. He coined the term *asymmetric carbon* for this condition, although the symmetry in question was not a property of the carbon atom but of its environment.[78] Armed with this model he could now assert that an isomer would possess optical activity if its composition included at least one asymmetric carbon. Verifying this hypothesis could be done by using a substitution reaction to replace the lone hydrogen in an optically active isomer, with either a hydroxyl, carboxyl, or methyl radical, and then checking that this led to the loss of rotatory power.[79]

The tetrahedral model itself, on the other hand, did not provide a causal explanation for optical isomerism, not did it assign a causal role to spatial distribution.[80] Those who rejected the existence of invisible atoms were therefore free to accept the model for its predictive power. But some chemists did try to find correlations between spatial properties, such as *the angle formed by two bonds*, and chemical behavior. One such chemist was Adolf von Baeyer, who in 1885 explored the causal role of bond angles using both Kekulé's closed ring model and the version that van't Hoff's gave of it, in which carbon-to-carbon single bonds were represented by tetrahedra sharing vertices, while double bonds involved the sharing of edges. In the tetrahedral model the four bonds join in the center at an angle of 109.5°, but when the carbons form a closed ring this angle can become deformed or *strained*. By manipulating mechanical models using springs, von Baeyer calculated that the most stable cyclic configuration, the one with the least amount of strain, contained five carbons. Configuration with a greater or lesser number would become deformed, and would therefore contain energy of deformation, a surplus of energy that could explain why organic substances like diacetylene exploded when heated.[81] Von Baeyer's hypothesis was correct only for planar arrangements of carbons, but despite its limited scope the strain hypothesis increased the plausibility of considering spatial arrangement as a true causal factor in the determination of chemical properties.

Spatial structure also became important outside of iconic models, as laboratory practices began to catch up with modeling techniques. In particular, synthesis ceased to be merely a way of validating the results of analysis, and became an exploratory tool in its own right, creating the expectation that carbon chains and rings could be assembled artificially. Prior to this period, at least one organic substance had been synthesized, urea in 1828, an achievement that had already shown that the bodies of animals and plants were not the only laboratories capable of producing complex carbon compounds. But the 1860s witnessed the synthesis of so many artificial substances that the French chemist Marcellin Berthelot was able to declare that chemistry created its own subject.[82] Berthelot outlined an ambitious project of *total synthesis* in which the chemist would start with carbon and hydrogen to yield the elementary building blocks. Using these in combination with oxygen would then lead to the synthesis of alcohols. Alcohols, in turn, could be used in combination with acids to create ethers; combined with ammonia to create substances like morphine, quinine, or nicotine; and with further oxygen to yield aldehydes and other aromatics.[83] Berthelot himself was able to carry out only a small part of this program, but others would continue it into the twentieth century, making the ideal of total synthesis a reality. Among the substances that he managed to create in his laboratory there were *polymers*, substances made of repeated copies of the same molecule. The name was coined by Berzelius, but it was Berthelot who delivered the first lecture on the subject in 1863, a lecture in which he hypothesized that if atoms of hydrogen or chlorine could be added to a molecule, then chemical reactions could be found that added further molecules identical to it. This implied that any unsaturated alcohol or acid should be able to polymerize.[84] At that time he had been able to synthesize only a few polymers, like pinene and pentene, but he went on to create polyethylene and propylene, using both heat and catalysts to drive polymerization.[85]

The German chemist Emil Fischer combined in his own personal practice van't Hoff's ability to use hypothetical models to reason about isomers with Berthelot's prowess as a synthetic chemist. He used these joint abilities to investigate the portion of the organic domain constituted by *sugars*. From van't Hoff, Fischer took a formula relating the number of possible isomers to the number of asymmetric carbons:

$$n \text{ asymmetric carbons} = 2^n \text{ isomers.}[86]$$

This formula predicted that sugars like glucose and fructose, each with four asymmetric carbons, had to have 16 isomeric variants. By 1894, Fischer had assigned structural configurations to 11 of these possible isomers.[87] This achievement crowned a ten-year-long research program that yielded a classification of all monosaccharides, producing a complex network of substances linked by chemical reactions.[88] His methods and results were uncontroversial because, as an anti-atomist, he relied more on empirical evidence about genetic relations between actual substances than on hypothetical spatial distributions of affinities. To him, the model of the tetrahedron was nothing but a useful isomer-counting device. But Fischer did venture some important causal hypotheses of his own. He had routinely used yeasts as diagnostic tools for his research on sugars and was able to observe that yeasts had a preference for one of two isomers that were mirror images of each other. This led him to think that the yeast's enzymes and his sugars fitted each other like *a lock and a key*, the key fitting only the isomer with the appropriate handedness, an insightful analogy still useful today.[89]

Let's assess the impact that these personal practices had on the consensus practice of 1900, using a textbook as reference.[90] The main difference with the previous textbook is that the concept of *structure or constitution* has become accepted, although it is still compatible with different ontological stances, referring either to the connectivity between macro components according to their valencies or to their micro spatial arrangement in a molecule.[91] Several concepts related to the concept of structure have also become consensual: *isomeric substances, polymeric substances*, and *tautomeric substances*. The existence of isomers was already part of the consensus in 1860, but they are now classified into three types: those with dissimilarities in their carbon chains; those with differences in the position of substituents in the carbon chain; and those with different spatial relations between elements and radicals.[92] A more novel addition is that of polymers, substances with the same components and proportions but *different* molecular weight. This condition can be illustrated by a substance like amylene, or C_5H_{10}. Polymers of amylene can have different overall weight because they contain different numbers of the same module, like $C_{10}H_{20}$, $C_{15}H_{30}$, and $C_{20}H_{40}$.[93] Tautomers are substances like ethyl aceto-acetate which behave in chemical reactions as if they possessed two different structures, making it difficult to assign them a structural formula. A variety of explanations are given in this textbook for tautomerism—they are isomers that interconvert reversibly into one another; or mixtures of two distinct substances coexisting at equilibrium— proof of their continuing problematic status.[94]

The concept of a carbon chain has now been fully integrated and diversified: not only are the *open-chains* formed by substances like methane distinguished from the *rings* of benzene, but these two basic structures are conceived as giving rise to more elaborate ones. Open-chains can have a linear form with or without branching, while rings are capable of having a variety of *side chains* attached to them. Moreover, it is now accepted that a ring need not be made exclusively out of carbon, but that it can contain a variety of other elements.[95] All these concepts are, in turn, used to explain regularities in isomeric substances, with different isomers arising from the presence or absence of branches, or from differences in the side chains. The specific case of isomers with different optical behavior is firmly linked to van't Hoff's ideas, making the concept of an *asymmetric carbon* part of the consensus. Moreover, more speculative ideas due to the same chemist—that single bonds allow rotations of carbon atoms while double bonds restrict them—are used to explain the difference between the isomers of saturated substances like lactic acid, and those of unsaturated substances like maleic and fumaric acids.[96] Finally, the idea that bonds may be strained when placed in tension, and that this deformation is causally related to their stability, is acknowledged.[97]

A variety of general statements shaped consensus practice in 1900. Some of these statements concerned the number of possible isomers that a substance may have, evidence for which was produced by syntheses of the isomeric variants, and by the use of structural formulas to explore the different possibilities. For a substance like propane, C_3H_8, in which components are connected like this—CH_3–CH_2–CH_3—the following statement was believed to be true:

1 Propane can have two isomers depending on whether a central hydrogen or a terminal hydrogen is replaced by a methyl radical.[98]

More general statements could be produced by reasoning along similar lines about entire families of open-chain substances, like the paraffins:

2 The number of possible paraffin isomers increases rapidly with the number of carbon atoms. There are three for C_5 (pentanes) and five for C_6 (hexanes), but for C_{13} there may be as many as 802.[99]

The degree to which evidence for statements like these could be produced in the laboratory varied with the complexity of the substance. Thus, all five isomeric variants of hexanes had been isolated, as had five out of the nine possible isomers for heptanes. But the larger number of isomers associated with substances with 13 carbons was entirely hypothetical. On the other hand, from the instrumentally confirmed statements, many others could be produced for families of substances that could be obtained in the laboratory from the paraffins. Thus, the textbook asserts that propyl alcohol, C_3H_7–OH, can have two variants; butyl alcohol can have four; and amyl alcohol can have eight. More generally:

3 The number of possible isomers for alcohols can be derived from the formula of the corresponding paraffin by determining in how many different positions the OH radical can be introduced.[100]

Serial classification schemes have been greatly improved. We saw that the consensus of 1860 already included a well worked out series of carbon compounds, the paraffin series, which could be expressed in compact form as C_nH_{2n+2}. This formula expresses the idea that as the number of carbons, n, increases, the number of affinity units becomes 4n, but because some of these are used to bind carbons into a chain, two for every pair, the number available to bind hydrogens is $4n - (2n -2) = 2n +2$. By 1900, many more series like these are known: first, there are the series of saturated and unsaturated alcohols, the formulas for which are $C_nH_{2n+1}OH$ and $C_nH_{2n-3}OH$, respectively; then, the series of saturated ethers, expressed as $C_nH_{2n+2}O$; the series of aldehydes and ketones, with the formula $C_nH_{2n}O$; and the series of saturated acids, $C_nH_{2n}O2$.[101] The textbook also shows how these series can give rise to further series by substituting hydrogens with other elements. Thus, each member of the paraffin series can give rise to series of halogen derivatives, by using chlorine, bromine, or iodine as substituents.[102]

This extended serial taxonomy improved the degree of order of the organic domain. Nevertheless, the validity of each of the series depended crucially on the correct assignment of valency number to elementary substances. Hence, any phenomenon that made this valency assignment problematic raised important questions. If valency was conceived as a property, for example, it seemed necessary to think of it as fixed: carbon should always be tetravalent and oxygen divalent. But as discussed in the previous section of this chapter, in inorganic chemistry this fixity could be called into question because metals formed compounds in ways that displayed *multiple valencies*. At first organic chemists could dismiss this problematic behavior, but eventually they were forced to confront it as certain organic compounds were found in which oxygen seemed to be tetravalent.[103] Thus, by 1900 the following problem had become part of the consensus: Why is a property like valency variable rather than fixed?

The solution to this general problem would have to wait for the creation of an electronic model of valency, but meanwhile particular cases of multiple valency raised more concrete problems. Thus, after mentioning that divalent sulphur can enter into compounds in which it behaves as tetravalent, the author raises a question with the following form: Why do sulphur compounds that should have property X if sulphur were divalent, have instead property Y as if it were tetravalent?[104]

Another problematic phenomenon first posed by metals, but later on found to have an organic counterpart, was *catalysis*: substances that can selectively influence a chemical reaction without themselves being transformed by it. The concept of catalysis was created by Berzelius in 1836, together with a hypothetical explanation for the phenomenon: the catalytic power of metals like platinum consisted in awakening dormant affinities in the reactants.[105] The practical importance of catalysts in industrial contexts kept the problem alive, but it did not strike organic chemists as something demanding explanation until *fermentation* forced it on them. This reaction, in which sugar is decomposed into alcohol and carbon dioxide, was traditionally carried out using yeast as a ferment, but a debate raged until the late 1850s about whether the relevant causal factor was whole micro-organisms or only a chemical component of them. When it was shown that crushed yeast cells retained fermenting power, the debate was eventually settled. But it was the isolation of the first enzyme (zymase) in 1897 that made catalysis into a urgent problem for organic chemists.[106] As in the previous case there was a general problem and a host of more concrete ones, addressing the selectivity of catalysts or their specific synthetic powers:

Why do some substances selectively affect the course of a reaction without themselves being affected by it, instead of being transformed by it like the other reactants?

Why does an enzyme that ferments a sugar (e.g. glucose) not ferment its isomer, instead of fermenting both?[107]

Why do small quantities of some acids (e.g. hydrochloric and sulphuric acid) catalyze polymerization in aldehydes?[108]

Why do substances that separately cannot induce a transformation between two isomers do so in combination?[109]

Finally, the consensus of 1900 also included the reasoning patterns that we have reconstructed using explanatory schemas. In the previous section we saw how changes in these patterns could be represented by suitable changes in the Affinity schema. There were at least two separate variants because not everyone agreed that affinity forces were quantized, and because the problem of variable valency was still unsolved. Changes in reasoning patterns about composition can also be tracked that way. At this point in time, most chemists agreed that arrangement in space made a difference in the explanation of the properties of a whole. Thus, the Part-to-Whole schema can be said to have been updated to this:

Question
Why does substance X have properties Y and Z instead of other properties?

Answer
PW-1. Properties are explained by the Part-to-Whole relationship between a substance and its components.

PW-2'. The properties of a whole are a blend of the properties of its components in the case of mixtures, or a set of properties different from those of its components in the case of compounds.

PW-3. The most basic components are those that cannot be further decomposed using existing chemical operations.

PW-4'. The number of basic components is determined empirically.

PW-5". The properties of a whole are determined by the nature of its components, by their proportions, and by their arrangement in space.

3 PHYSICAL CHEMISTRY

The Hybridization of Cognitive Tools

For most of the nineteenth century, organic chemists were focused on problems about the composition of substances. The other component of the domain, chemical reactions, was relatively neglected, playing the role of an instrument to generate series of kindred substances. So it was in *physical chemistry*—a specialized subfield that developed alongside its organic counterpart, stabilizing its identity only in the 1880s—that chemical reactions became phenomena in need of explanation. Several changes to the existing repertoire of cognitive tools had to be made to make this possible: new concepts had to be introduced, while much older conceptual distinctions, like that between compounds and mixtures, had to be refined; a large number of quantitative statements expressing the results of measurement operations of newly discovered properties had to be produced; numerical regularities in this data had to be found; novel problems had to be framed; and finally, *mathematical models* had to be borrowed from neighboring fields and adapted to the problems posed by chemical reactions. Let's begin by examining the conceptual changes involved.

As we saw in Chapter 1, some chemists in the second half of the eighteenth century had attempted to incorporate transformations into the chemical domain as part of an effort to elucidate the concept of affinity. Displacement reactions, then the exemplary case of a chemical reaction, gave rise to two explanatory models. One model postulated that when a neutral salt, composed of an acid and a base, reacted with an acid that had a greater affinity for its base, the second acid completely displaced the first one. Because the transformation was therefore *fully completed*, its products possessed a clear and distinct identity: a new salt and the displaced acid.

The other model postulated that affinity was not the only force affecting the substances in a reaction: temperature, concentration, volatility, and saturation effects also affected the final outcome. Thus, the tendency of a substance to react with another could decrease as the reaction proceeded and one substance saturated the other. Similarly, any factor that interfered with the intimate contact between particles needed for affinity to work, factors like volatility or cohesion, could weaken or even reverse its effects. Given this complexity, it was conceivable that a chemical reaction was in fact *incomplete*, yielding its regular products as well as uncombined portions of the original reactants. Thus, when two salts reacted in a double decomposition reaction, the product might be not two new salts with their bases and acids fully exchanged, but all four possible combinations coexisting in a mixture.[1] Physical chemistry was born as part of an effort to develop the insights contained in the second model, a development demanding the improvement of all the concepts involved, through modification, clarification, and differentiation.

The first concept was that of *chemical equilibrium*, the state at the end of a displacement reaction. Originally, the state of equilibrium was conceived as a state in which the balance of forces prevented any further change. This static conception had to be modified into the notion of a dynamic equilibrium, in which changes continued to occur but in such a way that the products of the forward reaction were counteracted by those of a reverse reaction. Then the concept of a *reversible reaction* had to be developed through a clarification of the relations between affinity and the other forces affecting the course of a chemical transformation: if the only relevant force was affinity then its selectivity should determine a unique direction for a reaction, but if other forces were involved then it was conceivable that a reaction might occur in both directions. Finally, the concept of a *mixture* had to become fully differentiated from that of a compound. When the original version of the second model was first introduced, a confusion between compounds and mixtures had led its creator to deny that compound substances displayed definite proportions of their components. This, of course, went against all the stoichiometric evidence that was accumulating at the time, but it was not unreasonable given that mixtures, such as a salt dissolved in water, exhibit indefinite proportions.[2] The apparent incompatibility of the model and the results of stoichiometry led to its neglect for the next 50 years.[3]

In addition to these conceptual changes, quantitative statements had to be produced in two areas of experimental chemistry operating at the frontier with physics: electrochemistry and thermochemistry. The latter was born in the late eighteenth century alongside the calorimeter, an instrument with which the heat generated during chemical reactions could be measured. Using more accurate instrumentation, many *heats of reaction* were measured in the following century leading to the discovery that the same quantity of heat was produced regardless of the path that the reaction followed: a transformation that took two steps to produce a given product yielded the same heat as one that took five steps to yield the same product. This discovery anticipated the physicists' idea that the amount of energy in a system is a function of its state at any one moment and not of the process that brought it to that state.[4] The dependence of heats of reaction on temperature was fully explored in the second half of the century, when several thousand calorimetric measurements were performed on a variety of chemical processes: the neutralization reaction of acids and bases; reactions of compounds of nonmetallic elements; the solution of substances in water. The data on heats of neutralization, heats of formation, heats of solution, and heats of dilution not only became a valuable addition to the reservoir of quantitative statements, but it also led to the discovery of significant regularities in the relations between the measured properties.[5]

The other source of novel data, electrochemistry, was born at the turn of the century when continuous electrical current became available thanks to the invention of the electric battery. The ability of electricity to break down compounds made it a valuable analytical instrument, but it was the realization that the conduction of electricity through a liquid was a chemical process that made it into a phenomenon. When thinking about the original battery, made of alternating zinc and copper discs with an interposed liquid solution of sulphuric acid, physicists assumed that the ability of the device to produce a continuous current was explained by the contact between the two dissimilar metals, the interposed liquid acting as an imperfect conductor, allowing electrical induction to take place. But when chemists integrated the device into their experiments they showed that the crucial component was the liquid solution itself: it was the series of *decompositions and recompositions* undergone by the dissolved chemical substance (the electrolyte) that actively produced the electrical current. The metal components were indeed important, the more oxidizable one becoming the positive pole of the circuit and the less oxidizable the negative one. But without chemical changes in the liquid, no current was produced.[6] Measuring the properties involved in the electrochemical process, therefore, was likely to yield insights useful to solve the problems posed by chemical reactions.

The first property that attracted attention was the relative amounts of decomposed substances (ions) that gathered at the poles—oxygen and hydrogen in the case of water, for example—and their dependence on the quantity of electricity used for decomposition. After experimenting with many electrolytes, a surprising conclusion was reached: the amount of ions created by a given quantity of electricity was proportional to the equivalent weights of the substances.[7] This was a significant regularity that demanded explanation. Measurements of a second property, the relative speed with which the ions migrated to the poles, revealed that for a large variety of electrolytes the migration speeds were different for the negatively and positively charged ions. This led to the postulation of another regularity: the relative migration speeds were independent of the strength of the current but varied with the initial concentrations of the electrolytes.[8] A third set of quantitative statements was produced by measuring the relative electrical resistance of different electrolytes. This showed not only that each substance had a definite and constant amount of resistivity, but from these numbers the conductivity of the solution could be calculated, conductivity being the reciprocal of resistivity. An interesting regularity was found by studying how conductivity varied at constant temperature for different concentrations: the conductivity of an electrolyte in solution *increased* as its concentration *decreased*.[9]

These results from thermochemistry and electrochemistry, in turn, posed problems that practitioners had to confront:

Why does the amount of heat evolved in a reaction depend not on its path but only its initial and final states?
Why are the amounts of substances deposited at the electrodes proportional to the substances' equivalent weights?
Why does the speed of ion migration towards the electrodes differ for differently charged substances?
Why does the electrical conductivity of an electrolyte increase as it becomes more diluted?
Why does the amount of a dissolved substance lower the freezing point of solutions regardless of its chemical identity?[10]

When these and other questions were originally asked, none were felt to be particularly urgent. But as they accumulated, the need for new cognitive tools to frame the problems posed by the existence of these *quantitative dependencies* among measured properties became obvious to a few practitioners. Mathematics had been recruited centuries before by physicists confronting similar problematic phenomena, and their exemplary achievements could serve as a guide to apply mathematical models in the field of chemistry. However, most chemists did not have any mathematical training for most of the 1800s, and most of them were highly suspicious of the severe idealizations needed to create models of chemical reactions. To understand the difficulties faced by chemists, let's first discuss the nature of these cognitive tools. A mathematical model consists of two components: *an equation* in which dependencies among properties are represented as relations between variables; and an *ideal phenomenon* serving as the referent of the equation.

An ideal phenomenon can be conceived as the result of a rigorous conceptual procedure, akin to a thought experiment: imagine a balloon filled with a gas like oxygen; then imagine an operation (like that of a pump) that when applied to the balloon reduces the density of its gas by half; repeat the operation many times to generate a series of balloons of decreasing gas density and then imagine *the limit of that series*. The end result is a well-defined imaginary phenomenon, *the ideal gas,* which played an exemplary role in the transmission of modeling techniques from physicists to chemists. In order for an ideal phenomenon to pose a problem, the way in which its laboratory counterparts do, *relations of statistical relevance* have to be established between its properties. These properties are, in the case of the ideal gas, temperature, pressure, and volume. The job of the equation is to describe these relations, not to explain them: the equation may correctly state that changes in pressure *strongly correlate* with changes in volume, but it does not explain why. The explanation may be, for example, that when a piston moves inside an engine it decreases the volume within a chamber, thereby increasing the pressure exerted by the steam inside that chamber, but the equation is silent about this causal process. But if the regularities captured by the equation are real—if an approximation to an extremely low-density gas is created in the laboratory and the measurements display the required statistical correlations—then the ideal gas *does pose a problem*, the problem of explaining why these regular relations characterize the ideal phenomenon.[11]

Chemists were not unfamiliar with idealizations because the use of Berzelian formulas demanded the rounding up of messy measurements into neat integer numbers. And the use of formula equations to model reactions also had to disregard details, such as the production of certain byproducts. But in the case of mathematical models, idealization went much further: the dependencies captured by the ideal gas equation were not those existing between the properties of a laboratory phenomenon but those possessed by a formal entity without independent existence. Moreover, the idealization ignored the chemistry of gases: at the limit of low density, a gas contains such few molecules that their interactions can be ignored, so its chemical nature becomes irrelevant. As the density is increased, however, not only the behavior of pressure or temperature deviates from the predictions of the model, but the deviations themselves depend on the chemical nature of the gas: water or carbon dioxide, for example, display larger deviations than nitrogen or hydrogen.[12] Finally, the lack of trust in the cognitive value of mathematical models may have been caused by their relation to laboratory evidence. The traditional way of linking models and evidence was to discover a real phenomenon, such as a new substance with intriguing properties, and then build a simplified model of its composition to explain its properties. The ideal gas model *reversed this familiar order* since the model was constructed first, and a laboratory phenomenon was created later to approximate the ideal: an artificially low-density gas. When the subfield of physical chemistry had become stabilized, its practitioners had become accustomed to this reversal: they did not see the laboratory as a place where analysis and synthesis were performed, producing new elementary or compound substances with properties that had to be explained, but as a place where measurements of highly simplified phenomena were carried out to test the validity of a mathematical model.[13]

Physical chemists were convinced that breaking with tradition was worth the risk because of the track record of mathematical models in physics. In particular, the history of this field seemed to show that once general statements about the dependencies between measured properties had been casted mathematically, equations could be combined with one another to generate new equations. This ability to serve as the nucleus around which many regularities could be interconnected—the nucleus around which *families of related models* could be generated—is what made equations useful despite the unrealistic nature of the phenomena being modeled. Let's look in more detail at how the model of the ideal gas became the nucleus around which *thermodynamic* models—models involving energy conservation, transformation, and degradation—were developed. The original ideal gas equation captured the regular variation in the volume of a gas as its pressure changed and its temperature was kept constant. This equation was combined, 140 years later, with another one depicting the regular way in which volume changed with temperature, keeping pressure constant.[14] The resulting combined equation captured the way in which an ideal gas behaves when both temperature and pressure are changed. In other words, it captured *the state* of an ideal gas at any time. The equation of the state of a gas at the limit of low density became: $PV = BT$.

In this famous equation, T represents the absolute temperature, P the pressure, V the volume, and B is a constant calculated by measuring the density of a gas in conditions approximating the ideal. To connect this basic equation of state to thermodynamics, three steps were required.

First, additional properties defining the state of a gas had to be added: its amount of internal energy and the degree of quality of this energy, that is, the entropy of the gas. Second, the equation had to be used to model the simplest state of a gas: *its state of equilibrium*, the state at which its entropy is at a maximum, while the values for temperature and pressure are uniform throughout the gas.[15] Third, a conceptual phenomenon, the *ideal engine*, had to be created to be the referent of the model. In the real steam engines used in the nineteenth century, temperature differences between a hot boiler and a cold condenser generated a spontaneous flow of energy from the boiler to the condenser, a flow that was converted into mechanical work through the action of steam on a piston. Because these real engines worked only for as long as the temperature differences were maintained, the conversion of heat into work took place *away from equilibrium*. This non-equilibrium state could not be modeled using the mathematics available at the time, but a more accessible version of the engine could be created through idealization. In the ideal engine only the transfer of energy from the boiler to the fluid and from the fluid to the condenser were included—all other transfers, like that from the fluid to the piston, were eliminated—and the transfer of energy was imagined to occur in extremely small increments, with the piston moving at extremely slow speeds.[16] This allowed physicists to treat the cyclic changes in the ideal engine as *infinitely small displacements from equilibrium*.[17]

Physical chemists could benefit from this cognitive resource but only by inventing chemical analogs for both the equations and the ideal phenomena. Since the eighteenth century, the majority of chemical reactions performed in laboratories took place in liquid solution, so an obvious analog for the ideal gas was a solution in which the substance dissolved (the solute) was present at extremely low concentrations. A counterpart to the equation of state could then be developed if chemists could show that the properties involved—absolute temperature, pressure, and volume—displayed the same dependencies in liquid solutions at the limit of low concentration.[18] Keeping the temperature constant, the main dependency between properties that had to be captured was that between pressure and volume, that is, chemists had to show that pressure changes were inversely proportional to volume changes. A good source of data to check whether this dependency actually existed in liquid solutions was the phenomenon of *osmosis*: the capacity of water and other solvents to pass through a membrane. In the case of selective membranes, those that allowed the solvent but not the solute to pass through, the pressure that the dissolved substance exerted on the membrane could be measured by applying an external pressure to the solution until the flow of solvent stopped. The resulting value for this *osmotic pressure* was shown to be directly proportional to the concentration of the solute, and hence, inversely proportional to the volume of the solution.[19]

This was precisely the dependency of volume on pressure needed to show that extremely dilute solutions behaved just like extremely low-density gases. The resulting equation turned out to be identical to that for the ideal gas, including the value for B, the constant of proportionality. Then, as a second step, the new equation of state had to be connected to thermodynamics by developing a chemical analog of the ideal engine. This was achieved by imagining a vessel subdivided using a perfectly selective membrane, a membrane that did not allow the passage of any particle of the solute. One compartment was imagined filled with pure solvent and the other with the solvent mixed with a dissolved substance, such as salt. These two compartments were the analog of the boiler and the condenser. Like the two components of a steam engine, which embodied a temperature difference, the two compartments were thought of as embodying *an affinity difference*, a difference in the attraction of water for salt in each compartment. And much as a temperature gradient could be used to perform work on a piston, a gradient of affinity could be used to perform the work necessary to drive osmosis.[20] If we further imagined that osmosis was reversible, so that it could be carried on in a cyclic way, the dynamics of the entire system could be modeled in the same way as the ideal engine, using the smallest possible displacements from equilibrium to follow the states of the system as it went through its cycle. Adapting this quasi-static cyclic approach to chemistry involved creativity on the part of chemists, as did its extension to other forms of chemical interaction, such as those taking place during electrolysis.[21]

Let's summarize what has been said so far. A series of changes in the cognitive tools available to nineteenth-century chemists had to occur before the efforts of their predecessors to make chemical reactions into a phenomenon could bear fruit: concepts had to be refined and clarified; a large number of measurements had to be carried out on the properties of reactions (the heat produced, the rate of ion migration, the changes in concentrations of the reactants and products) and regular dependencies among the resulting numbers had to be discovered; mathematics had to be recruited as a resource to make sense of these dependencies and to predict new ones; and, finally, general statements from thermodynamics had to be integrated into chemistry. The two basic statements were that *energy is conserved through its many transformations* and that *energy transformations have a direction*, going from high-quality to low-quality energy. Evidence for the first statement had been produced when heat was found to be transformable into mechanical work, and when careful measurements had revealed that there was a mechanical equivalent of heat.[22] This statement helped chemists understand why all chemical reactions involved energy changes, but it did not provide them with any guidance regarding the pathway that a reaction could take. The second basic statement, asserting that the pathway always goes from fresh to exhausted gradients, supplied that guidance.[23]

The incorporation of thermodynamics into chemistry led to a deep conceptual change: it allowed practitioners to think of affinity not as a force but as a kind of energy. The name for the concept remained in flux for a while (work of affinity, free energy, chemical potential), but the concept was firmly established early on: a definite amount of work must be performed during any chemical reaction and this capacity to perform work depends on the existence of a gradient of chemical potential. As long as affinities were unequal in adjacent parts of a system, a flow of matter would tend to occur from one part to another until the differences had been equalized, or the gradient cancelled.[24] In addition to this, there were other conceptual obstacles to be surmounted, three of which were mentioned at the beginning of this chapter: thinking about reactions as always reaching full completion; thinking about equilibrium as a state in which nothing happens; and distinguishing compounds from mixtures. The three difficulties were interrelated: if equilibrium was thought as a dynamic state in which reactions in opposite directions took place at the same time, it became easier to understand why the end state of a reaction might not be a pure final product, but a mixture of all possible products. In other words, if the reaction of two substances, A and B, was complete, then the end state should be a pure compound AB. But if the reaction did not reach completion then its final state would be a dynamic equilibrium in which the reaction A + B = AB took place alongside the reverse one, AB = A + B, and the final product would be a mixture of A, B, and AB.

Let's begin the discussion of how these conceptual obstacles were overcome with the question of how chemists learned to tell compounds and mixtures apart. In this task they were aided by another cognitive tool imported from physics: *phase diagrams*.[25] By the end of the century, it was well established that there was a causal connection between phase transitions—like the transformation of a gas into a liquid or a liquid into a solid—and the complete or incomplete character of a reaction: if in a chemical reaction taking place in liquid solution one of the products changed phase, that is, if it escaped as a gas or it sedimented as a solid, that product was effectively removed from the transformation becoming unavailable to enter into the reverse reaction. Hence, a chemical reaction involving a phase change tended to be complete and to produce a pure substance as its final product.[26] But an even more intimate connection could be made between these concepts if the very distinction between a pure compound and a mixture was *conceived in terms of phases*, their coexistence in incomplete reactions conceived by analogy with a liquid and a gas coexisting when vaporization and condensation occur at the same rate in both directions. This was just an analogy but its value was that the relation of coexistence could be rigorously studied using phase diagrams.

A phase diagram is a geometric representation of the space formed by all possible combinations of temperatures and pressures achievable in the laboratory. If pressure is kept constant, the equilibrium state appears as a singular point in the temperature scale, but when both temperature and pressure are allowed to vary, the different equilibrium states appear in a diagram as a series of points, that is, as lines or curves along which different phases coexist: boiling point curves (in which the gas and the liquid state coexist); melting point curves (with coexisting liquid and solid states); and so on. The set of curves captures *the structure of the space of possible combinations* of temperature and pressure, or in the case of chemical reactions, of temperature and substance concentration. This allowed practitioners to apply geometrical reasoning to speculate about the range of possible outcomes of a chemical transformation. For example, when two curves met at an angle (forming a line with a kink) this indicated the presence of a pure compound possessing *discontinuous* properties, while two lines forming one continuous curve indicated the presence of a mixture with a blend of properties.[27] Moreover, *regions of stability* could be discerned in these diagrams. The inside of the region contained the combinations of values of temperature and concentration that yielded a pure compound, while those on the outside of the region led to the production of a mixture.[28]

The analogy between phases and the end states of a reaction could be carried further with the aid of mathematical models. There were two ways of approaching the modeling task: *macroscopic and microscopic.* The advantage of a macroscopic approach is that it focused on collective or bulk properties of a phenomenon, like temperature or concentration, so the validity of its models did not depend on any particular assumption about the nature of atoms or molecules. To counter this advantage, the proponents of the microscopic approach showed that an ideal gas could be modeled by making only a minimum number of assumptions: that the volume of the particles was so small that it could be neglected; that the particles collided but the duration of their collisions was small compared to the time elapsed between collisions; and that at extreme low densities the attractive forces between molecules could be ignored.[29] From the point of view of it influence on physical chemistry, the most important of these microscopic models targeted the partial evaporation of liquids in a closed vessel. In this model, some particles were assumed to move with enough velocity to overcome the forces of cohesion of the liquid, while simultaneously some of those liberated particles lost velocity and were captured again by the liquid. This dynamic progressively reached a limit at which evaporation and condensation happened simultaneously, while the overall vapor pressure (the effect of the particles colliding against the walls of the container) became constant.[30] A chemical analog of this model could be created by conceiving of the end state of a reaction as involving a similar dynamic: instead of particles in the gas and liquid state exchanging places at the same rate, transformed and untransformed substances had to be thought of as being produced at the same rate.[31] Once the analogy had been casted mathematically, all three conceptual barriers to a better understanding of the outcome of chemical reactions could be removed.

We are now in a position to judge whether the study of chemical transformations improved in the nineteenth century. That temperature, concentration, solubility, and other factors acted together with affinity to determine the outcome of a reaction had been long known, but the effect of these other factors was expressed in relatively vague terms. Phase diagrams, on the other hand, provided detailed maps of the possibilities open to a chemical reaction, in which the effect of every combination of factors was clearly indicated. This was a definite improvement. So was the clarification of the distinction between complete and incomplete reactions, as well as the new way of understanding the final outcome of incomplete reactions as a state of dynamic equilibrium. Thermodynamic concepts (energy, entropy) and basic statements about the conservation of the former and the tendency of the latter to either increase or stay constant led to an improved grasp of the changes that took place during a chemical reaction, as well as the overall direction of those changes. In particular, the new conception of affinity as operating through gradients, like temperature or pressure, allowed chemists to fit affinity phenomena into a more general pattern: the direction of any process that involves energy changes is determined by the tendency of *all* gradients to cancel themselves. This change in the concept of affinity, on the other hand, captured only one part of the old concept, the attraction part of "selective attraction," while selectivity had to be accounted for using concepts like valency, as in the idea of the tetrahedral grasp of carbon. In a sense, the explanation of the selectivity had to follow the microscopic approach, in which commitment to the existence of invisible particles was necessary, while the explanation of the attraction could be made at a macroscopic level, using concepts like the work needed to break substances apart into their component substances, or the work needed to bring substances together into a new substance.

The incorporation of thermodynamic concepts and statements also led to an improved understanding of the composition of reactants and products. At the beginning of the century, heat was conceptualized as a material substance (caloric) that could enter into combinations with other chemical substances. Pure oxygen gas, for example, was viewed as a *compound* of oxygen and caloric. The only difference between heat and oxygen was that the former was imponderable, like light or magnetic fields. Now that the imponderables had been shown to be transformable into each other, they had become unified under the concept of energy.

But even if heat had stopped being conceived as a substance it could still be considered a component of a substance: *its energy content*. We can register this change in the way chemists reasoned about composition by making some changes to the Part-to-Whole schema, adding the clause PW-1a to statement PW-1, and changing PW-5" into PW-5"':

Question
Why does substance X have properties Y and Z instead of other properties?

Answer
PW-1. Properties are explained by the Part-to-Whole relationship between a substance and its components.

   PW-1a. The components of a substance are of two kinds: material substances and forms of energy.

PW-2'. The properties of a whole are a blend of the properties of its components in the case of mixtures, or a set of properties different from those of its components in the case of compounds.

PW-3. The most basic components are those that cannot be further decomposed using existing chemical operations.

PW-4'. The number of basic components is determined empirically.

PW-5"'. The properties of a whole are determined by the nature and spatial arrangement of its material components, and by the relative amounts of its material and energetic components.[32]

The properties of chemical substances in different phases are the most obvious example of properties in which both material and energetic components are important, the solid, liquid, and gas series displaying an increasing energy content. But another important example is electrolytic solutions. These mixtures of a liquid solvent and a liquid or solid solute became problematic in the 1880s because they defied the predictions of the mathematical model. When testing the chemical analog of the ideal gas, deviations from its predictions were expected because the very dilute solutions prepared in the laboratory were only an approximation of a solution at the limit of low dilution. But there were also unexpected deviations in the case of mixtures of a solvent and an electrolyte. When these solutions were diluted their osmotic pressures were consistently higher than they should have been. Because they violated the predictions in a regular way, this mismatch could be fixed by simply adding a correction factor: $PV = iBT$.

This maneuver was at first viewed as a makeshift solution, but soon other chemists discovered that the seemingly arbitrary constant did in fact have a referent. The need for the correction factor was explained by assuming that some of the solute's particles had become *dissociated into ions* merely as a result of their being dissolved, that is, without the need for an electric current.[33] The anomalous high values of osmotic pressure could then be explained by analogy with the case of a gas at very high temperatures. Chemists reasoned that just as the molecules making up a sample of oxygen gas break up into atoms at a high enough temperature, thereby increasing the amount of pressure they exercise on a container, molecules of salt dissolved in large quantities of water dissociated into ions of sodium and chlorine. This dissociation caused the solution to have twice the number of particles, and therefore be capable to exert twice the amount of pressure. This explained why the correction factor was $i = 2$ in the case of sodium chloride.[34]

Electrolytic solutions are a perfect illustration of a phenomenon explained through the Part-to-Whole schema, improved by the addition of energy content: the properties of the whole (the solution and solvent) depend crucially on energy content, that is, on the electrical gradient that is created when particles become dissociated into oppositely charged ions. These solutions are, in a sense, predisposed to conduct electricity, explaining why the migration of ions towards the electrodes is caused by even minute amounts of external current. Thus, it can be argued that the changes made to the Part-to-Whole schema were an improvement, not only in the case of phases—where the old caloric already "explained" the higher volume of gases—but also in the case of electrolysis. Finally, the mathematical models creatively adapted from physics represented an improvement in the way in which problems about chemical reactions could be posed: the costs of radical idealization were compensated by the way in which equations could give rise to entire families of other equations, linking together measurable properties in ways that could not be performed otherwise. Of all these improvements, the latter was the most controversial at the time, the one that took the longest to become part of consensus practice.[35] For this reason, whether the addition of mathematics to chemistry made the latter better is hard to assess if we confine ourselves to the evidence that was available to the actors at the time.

But as has been argued in this book, the merits of a cognitive tool should be judged not only by the way it affected contemporary practices but also by its (retrospectively assessed) track record. Because cognitive tools should be judged by their own local standards, we must begin by defining in what sense the newest addition to chemical practice could be made better. First of all, one way of improving a mathematical model is by the *relaxation of its idealized assumptions*. The ideal gas equation of state assumed particles with zero volume that did not interact with each other, so it could be made better by including interacting particles with a finite volume. This refinement was performed in small steps. At first it was too difficult to include multi-particle interactions, so models were created that included interactions between pairs of particles. This could be justified by the assumption that, from a macroscopic point of view, the density of the gas was still low enough that encounters between three or more particles were rare. The relaxation of idealized assumptions was not a routine maneuver and demanded creativity. It was not obvious, for example, how the variable V, referring to the volume of a gas as a whole, should be modified to include particles with their own volume. Could it be done simply by subtracting one from the other? If the variable b was used to represent the sum of the volumes of the particles, this changed the ideal gas equation to: $P(V - b) = BT$. A similarly simple change was to add the overall contribution of interactions as just another variable (ø) affecting the pressure: $(P + ø)(V - b) = BT$.

These simple changes, however, proved inadequate.[36] The variable b had to be calculated in a more sophisticated way, taking into account the average distance that a particle travelled between collisions (that is, its mean-free path), and the statistical distribution of particle velocities. And the variable, ø, the effect of particle interactions on pressure, had to be expressed as series a powers of the gas density.[37] As more terms from the series were included, the equation became more realistic. Conversely, truncating the series after the first term yielded the original equation as a limiting case. Various chemists used this improved version of the model, adapting it to the case of chemical reactions by adding one term for the mutual attraction of the solute molecules and another one for the attraction between solvent and solute.[38]

A second way in which mathematical models could be made better was by *extending them to non-equilibrium states*. The models adopted by physical chemists were useful to discover dependencies among the properties of chemical reactions when they had reached equilibrium, that is, when the entropy of a transformation had reached a maximum. This implied that only the initial and final states of a reaction were important, not the path followed from one to the other. When there was a need to model changes of state (as in the ideal engine), these were treated as if they occurred infinitely slowly. This was clearly a major limitation of the approach so its elimination constituted an improvement. Making the path to equilibrium part of a model involved the explicit representation of *the production of entropy* in a process, since this crucial property could not be assumed to remain constant at a maximum value.[39] In addition, while in the equilibrium case, differences in temperature, electrical potential, or chemical potential could be considered as effectively cancelled, these gradients (thermal, electrical, or chemical) and the flows they drove had to be included in non-equilibrium models. If the gradients were of low intensity they defined *near equilibrium* states in which the relationship between gradients and flows could be modeled as linear or proportional: small causes produced small effects, and large causes produced large effects. High-intensity gradients, on the other hand, characterized *far from equilibrium* states in which the causal relationship was nonlinear: small changes produced large effects and vice versa.[40]

Moving beyond the equilibrium state involved making changes to the two components of a mathematical model: the ideal phenomenon and the equations. In the near equilibrium case, the gas at the limit of low density had to be replaced by an *ensemble of small compartments* each containing an ideal gas. The compartments had to be large enough to have well-defined collective properties but small enough to be usable in the description of spatial variations of these properties. The new ideal phenomenon included gradients between compartments driving flows across them: flows of energy for thermal gradients, or flows of matter for gradients of chemical potential. This way the old mathematics could be preserved for the interior of the compartments while a new formalism was developed for the ensemble as a whole. Thus, we can consider the latter an improved extension or expansion of the former.[41] In the far from equilibrium case, the ideal phenomenon had to be transformed more drastically. To begin with, while near equilibrium the condition of the ideal gas is determined only by the current values of the variables in the equation of state, at greater distances from equilibrium *the history* of those variables has to be taken into account, as if the ideal material had a memory of the changes that its gradients had undergone.[42] In addition, while near equilibrium the flows driven by the gradients do not affect the equation of state, far from equilibrium they do, so new variables had to be included to *explicitly characterize those flows*. Thus, a richer ideal phenomenon was created, one in which flows of energy or matter could affect the amount of entropy through the history of the new variables.[43] Whether this more radical departure from ideal gases can be considered an improved extension or a radical innovation is still unclear.[44]

One way in which the second component of a mathematical model, the equations, had to be modified is obvious: the equations had to have variables to represent gradients and flows in one case, or the history of gradients and properties of the flows in the other. But a philosophically more important modification was that the mathematical behavior of the equations had to capture the *irreversibility* of the process of approaching equilibrium. In particular, the equations used to model the production of entropy had to be functions (such as Lyapunov functions) that *can only increase in time*.[45] The end state which such functions tend to approach varies for each ideal phenomenon: isolated systems, in which there are no flows of energy or matter, tend to move irreversibly towards a state of maximum entropy; closed systems, with flows of energy but not matter, tend towards a state of minimum free energy; and finally, open systems, with low-intensity flows of both matter and energy, have a tendency to move towards the state with a minimum entropy production. These three states are all stationary, *extremum* states approachable from a variety of initial conditions, so neither one has to incorporate history.[46] But to model the far from equilibrium case demands a different Lyapunov function, one in which the states that the ideal phenomenon irreversibly approaches need not be stationary or unique. In cases in which there are several possible steady states, accidents in the history of the system, such as an otherwise negligible fluctuation, can determine which of the available extremum states is picked, thereby introducing history into the model.[47]

It is too early to tell which of these ideal phenomena and modeling strategies have become part of the current consensus, or at least whether this consensual status is indeed merited. But what these new approaches do show is that the controversial models of the end of the nineteenth century were improvable, a piece of information that was not available to actors at the time, hence not usable to resolve their controversies, but one that is available to us to assess the value of the new cognitive tools. Given the creativity involved in the changes needed by both equations and ideal phenomena—creative changes that just as they occurred, could not have occurred—there is nothing teleological (or Whiggish) about this kind of retrospective assessment. Thus, we can conclude that the track record of mathematical models in the nineteenth century shows that these cognitive tools were a valuable addition to the chemist's repertoire, a clear improvement over the way in which problems about chemical reactions were posed.

From Personal to Consensus Practice 1800–1900

1800–50

John Dalton (1766–1844), Humphry Davy (1778–1829), Michael Faraday (1791–1867), Johann Döbereiner (1780–1849), Pierre Louis Dulong (1785–1838), Eilhard Mitscherlich (1794–1863), Germain Henri Hess (1802–50), Hermann Kopp (1817–92)

The domain of chemistry in the first decades of the nineteenth century was experiencing dramatic changes. Analytical techniques were undergoing rapid improvement and as a result the number of elementary substances was increasing enormously: from the 33 postulated at the end of the previous century, to the 54 known in the 1830s, to the 70 elements that would be classified into the Periodic Table in the late 1860s. The multiplication of substances constituting the limit of chemical analysis was not only philosophically jarring, conflicting with the long-held belief that the number of basic elements should be small, but also created difficulties for professors and textbook writers who had to transform a growing list of particular statements about the properties of elements, and of the compounds they could form, into a teachable discipline based on general principles.[1] The incorporeal or imponderable "substances," heat and electricity, continued to be the most problematic components of the domain. Phenomena of electrical and thermal nature inhabited two domains, those of chemistry and physics, and were therefore associated with different cognitive tools. Electricity, for example, was already being studied by physicists using mathematical models in the early part of the century, at a time when mathematics was not a part of chemistry.[2] But there was nevertheless traffic between the two domains. One type of transfer involved the conversion of what was a phenomenon in one domain into an instrument in the other. The best example of this was the Volta pile, an early form of the electric battery that in physics was used to display electricity as a phenomenon but that in chemistry was transformed into a powerful analytical tool, the very tool that had led to the proliferation of elementary substances.

Although in this period most members of the chemical community were exploring the organic portion of the domain, there were some who were attracted to the zone of contact with physics. This was the case of those who practiced electrochemistry, but it also involved chemists who were dedicated to the accurate measurement of a variety of properties which could be used to fix the referent of substance concepts, properties like crystal shape, specific heat, specific gravity, optical and magnetic behavior. Philosophers of science, particularly those who tend to reduce laboratory practice to the testing of predictions deduced from models, have never adequately conceptualized the activity of measurement, an activity that is often a goal in itself.[3] In the case of chemistry, one effect of the accumulation of numerical data about physical properties was to prepare the ground for the eventual reintroduction of chemical reactions into the domain. Although some eighteenth-century chemists (Bergman, Berthollet) had already approached transformations as objects of study, their research program was mostly abandoned by later practitioners, and its revival had to wait for the creation of the new subfield of physical chemistry in the 1860s. The work of those experimentalists, dedicated to measuring properties like the heat produced during chemical reactions, helped to change the status of reactions from instruments to phenomena.

The personal practice of the British chemist John Dalton began at the physical frontier of the chemical domain. Early in his career, Dalton became intrigued by the problem of why elementary substances like oxygen, nitrogen, and carbon dioxide do not form compounds through chemical reactions in the air. And if these substances simply coexisted physically, why did they not become stratified by weight? He published an essay in 1801 in which he speculated about these problems. One way to solve the second problem was to imagine substances as composed of self-repulsive particles: if particles of oxygen repelled only other oxygen particles, while nitrogen ones repelled other nitrogens, the entire particle population would not sort itself out into layers but form a homogenous mix. The capacity for self-repulsion, in turn, could be accounted for by postulating that the particles were surrounded by an atmosphere of heat. Since it was already part of consensus practice that the amount of heat (caloric) accounted for the differences between solid, liquid, and gas states—the greater the heat content the weakest the cohesion between components of a substance—this hypothesis was entirely plausible. The first problem could then be solved by postulating that the particles of different elements varied in size, each occupying a different volume, and that their heat atmospheres varied in radius. This variation in size would prevent the particle population from reaching equilibrium, creating the conditions under which they would ignore one another chemically.[4]

These two hypotheses were entirely physical, but Dalton's thoughts about invisible particles were also influenced by considerations from chemistry. He knew about stoichiometry, and that its practitioners had shown that knowing how much of an acid converts a given amount of base into a neutral salt gave information about the proportions of acid and base in that salt. He also knew that these proportions often displayed discontinuities, suggesting that the substances themselves were made of discontinuous atoms. This was also an entirely plausible hypothesis, but to make his case Dalton would have to find a way to bridge the cognitive gap between the macroscopic world of substances and the microscopic world of invisible particles.[5] Stoichiometry provided the bridge: given a particular compound for which the parts per weight of each component were known—water was made of 87.4 parts of oxygen and 12.6 parts of hydrogen, for example—the weights of the individual atoms could be calculated by making assumptions about the relative proportions of the components. Dalton assumed that if only one compound was known its proportions should be the simplest possible, that is, 1:1. If more than one compound existed, then the next simplest relations should be postulated. In the case of oxygen and hydrogen, only one compound was known to him, so he assumed that the ratio of their atomic weights should be 12.6 to 87.4, or expressed using hydrogen as a unit, that the weight of a single oxygen atom should be 7. Although this conclusion turned out to be wrong—the component proportions in water are really 1:2—this bridge between the macro and the micro scales served as an inspiration for the chemists who struggled with atomic weights for the next 50 years.

Another corner of the chemical domain also bordering with physics housed the personal practice of Dalton's fellow countryman, Humphry Davy. Like many European chemists at the turn of the century, Davy's practice was affected by the sudden availability of continuous electrical current. The Volta pile made its mark in the chemical community almost immediately by demonstrating the ease with which its current could be used to decompose water. Davy built larger and more powerful versions of this instrument and used it in 1807 to isolate sodium, calcium, boron, potassium, strontium, and magnesium, a list to which he added chlorine, iodine, and bromine three years later. Although the electrical battery was merely an instrument, Davy correctly saw that its power to break apart compounds raised the question of its relation with the force that holds them together: affinity. He experimented with different substances and observed correlations between the affinity of substances for oxygen and the intensity of the current.[6] Other chemists had already noticed that the selectivity of affinity had a counterpart in electrolysis, a relation that was particularly evident in the decomposition of water: hydrogen and oxygen tended to selectively accumulate at different ends of the metallic wire used to inject the current. Davy repeated the experiments, confirming the attraction of the elements of water for the differently charged electrodes, and went on to perform studies of the migration of acids and alkalis to opposite poles in the decomposition of neutral salts. These phenomena strongly suggested an electrical explanation of affinity, the first version of which Davy published in 1806.[7]

The personal practice of Michael Faraday, Davy's disciple and one-time assistant, straddled both sides of the frontier. In the 1830s, Faraday discovered several phenomena (electromagnetic rotation and induction) that became an integral part of the domain of physics, and brought order to its domain by showing that the different forms of electricity known at the time (static, voltaic, animal, magnetic) were manifestations of one and the same force.[8] But he also made crucial contributions to chemistry by determining the role played by chemical reactions in electrolysis. He introduced the electrochemical vocabulary still in use today: *electrolyte*, the dissolved substance to be decomposed; *electrode*, the metal making contact with the solution; *ions*, the decomposed products; and *anode* and *cathode*, the poles through which electrical current enters and leaves the solution, completing the circuit. These were not just new labels for old concepts. Before Faraday, for example, the positive and negative poles were thought of as simply the place at which the decomposed substances gathered, not the entry and exit points for the current. Similarly, prior to his work the liquid solution was thought of as a passive conductor of a fluid electrical substance. In Faraday's reconceptualization, on the other hand, the dissolved electrolyte actively maintained the circulation of the current though a series of *local compositions and decompositions*, a series in which electricity acted not as a substance but as a directional force. This implied that electrolysis was a chemical, not a physical, phenomenon.[9]

In addition to introducing new concepts and explanations, Faraday discovered important empirical regularities in electrochemical reactions: he established that the quantity of electricity circulating in a solution was proportional to its capacity to decompose a chemical substance. And in 1834, after experimenting with several electrolytes—tin chloride, lead borate, hydrochloric acid—he discovered that substances were broken into their components in proportions that neatly coincided with chemical equivalents. Because the technique used to calculate equivalent weights (neutralization reactions) was originally used to measure affinity, these results convinced Faraday of the identity of electricity and affinity.[10] To be able to go beyond Davy's hypothesis, however, Faraday had to prove that electrical and chemical phenomena displayed *polarity*, and that the asymmetry of positive and negative poles was related to the selectivity of affinity forces. One line of research aimed at obtaining evidence for this hypothesis involved the study of crystals of different substances (arsenic, antimony, bismuth) and the discovery that when suspended horizontally on a magnetic field the crystals aligned themselves with the lines of force of the field. That is, they exhibited polarity.[11]

Crystal shape had been used since the eighteenth century as one physical property useful to identify the referent of substance terms. More recently, and thanks to a greatly improved instrument to measure crystal angles (the goniometer), a taxonomy of shapes had been developed, but mineralogists had made no effort to link the regularities displayed in this classification to those of chemical composition. That is, morphological similarities had not been linked to stoichiometric ones. Using Berzelian formulas the German chemist Eilhard Mitscherlich was finally able to establish a firm connection between the two. Mitscherlich discovered that salts composed of very different metallic bases had identical crystal shapes, suggesting that shape was related not so much to the chemical nature of the components of a substance but to the number and grouping of its combinatorial units. This phenomenon, named *isomorphism*, was then established by him to exist in the sodium, ammonium, and barium salts of these metals.[12] Further links to stoichiometry were created by studying the crystals generated from the combination of sulphuric acid with a variety of metals (copper, iron, cobalt, zinc) and establishing a relation between their shapes and the number of units of water that the crystals contained.[13] Once this empirical correlation was established, the existence of isomorphous crystals could be used to constrain the choice between alternative formulas for a compound, and as an indirect means to determine the combinatorial units of newly discovered elements.

Metallic crystals were also involved in another novel phenomenon that had intrigued Faraday, the phenomenon of *catalysis*: the facilitation of a chemical reaction by a substance that is itself not transformed in the process. The phenomenon was discovered by another German chemist, Johann Döbereiner, when studying the way in which platinum affected the chemical reaction between oxygen and hydrogen. But this was not his most important contribution to chemistry. The population explosion of elementary substances had created the urgent need to bring order to this portion of the domain. This need would be met in the late 1860s when the Periodic Table was first introduced, but some of the patterns displayed by this taxonomic schema had already been discovered by Döbereiner. As early as 1817 he noticed that the equivalent weights of some oxides—calcium, strontium, and barium oxides—displayed a clear numerical regularity: if the weights of the first and last substances were added and then divided by two, the result was the weight of the middle substance. In 1829 he added two more *triads*: chlorine, bromine, and iodine, and lithium, sodium, and potassium.[14] This recurring relation between equivalent weights was remarkable, but weight was ultimately a physical property, while Döbereiner wanted groupings characterized by similar chemical properties. The first three triads met this criterion, but other sets of three elements, such as carbon, nitrogen, and oxygen, did not: their equivalent weights had the right proportions but they were not chemically similar, so Döbereiner rejected them. In retrospect it is clear that the relations he discovered correspond to the vertical columns of the table, which group elements by similarity of properties, but fail to take into account the regular *repetition* of properties which is displayed by the horizontal rows.[15]

In addition to multiplying elementary substances and acting as an incentive to find patterns in their distribution, electrochemistry raised problems about the relation between electricity and chemical reactions. But electricity was not the only imponderable "substance" playing this role. In the hands of Germain Henri Hess, the Swiss-Russian chemist who coined the term *thermochemistry*, the heat produced in reactions became a serious object of study. In the 1840s, Hess found that reactions in which an acid neutralized a base, the heat generated depended only on the initial and final states of the process. In other words, if there existed several reactions through which a neutral salt could be produced, some involving a larger number of intermediate steps than others, the overall amount of heat produced was exactly the same. This empirical regularity was an early statement of the principle of conservation of energy.[16] The French chemist Pierre Louis Dulong (in collaboration with Alexis Petit) performed extensive measurements of specific heats, refraction indices, heats of combustion, and other properties, and also discovered regular patterns. In particular, in 1819 he found that the amount of heat needed to raise the temperature of a given quantity of a substance by a specified amount (its specific heat) yielded a nearly constant number for many substances when multiplied by their atomic weight.[17] As a final example of chemists who dedicated their personal practice to the unglamorous activity of measuring the physical properties of chemical substances, we should mention Herman Kopp, a German chemist who devised a method for the accurate determination of boiling and freezing points, seeking to find correlations between these critical values and the composition of substances in experiments conducted between the years 1842 and 1855. In the case of organic compounds which, as we saw in another chapter, form series, some of which vary regularly in increments of CH_2, he found that the boiling points decreased with the complexity of the substance, and that they approached a limiting value as the series progressed.[18]

What do contemporary textbooks have to say about the content of consensus practice at the frontier between chemistry and physics in the year 1850?[19] Fifty years earlier, textbooks included three forces thought to be active during chemical reactions, two of them attractive (affinity and cohesion) and one repulsive (heat). Adhesion, as a third attractive force, was also discussed. But it is clear that in 1800 the three attractive forces had not become *conceptually differentiated*: the author speculates that cohesion and affinity may be one and the same force, one acting between similar particles while the other operates between dissimilar ones.[20] Adhesion was thought to be a form of affinity, because some eighteenth-century chemists had used the adhesion between metals and quicksilver to measure selective attraction, and because the ordering of substances by their degree of adhesion corresponded to their order by affinities.[21] In the 1850 textbook these three forces have become fully distinct, and their roles in chemical reactions better defined.[22]

The role of cohesion in the generation of solids was recognized in 1800, as was the fact that cohesive forces can generate both regular or irregular solid arrangements, crystals or glasses. Early classifications of crystal shapes by the way in which they broke or cleaved in preferential directions were acknowledged in that text.[23] Fifty years later, the way in which crystals are classified by their shape has been greatly improved, as has the way in which the process of crystallization itself is conceived. Thus, it is now known that when cohesion and adhesion are balanced in a liquid, as in a saturated solution, very small causes (a slight vibration, a small impurity) can tilt the balance in favor of cohesion; and that when a seed crystal is dropped into a saturated liquid, the particles' adhesion to the liquid can be overcome by the addition of solid surfaces to which they can adhere instead.[24] These improvements were accompanied by the introduction of novel concepts, such as the concept of isomorphism. The convergence of geometry and composition exhibited by *compound* substances was problematic because the elements that composed them tended to exhibit different crystal shapes. Once these elements entered into combination, however, the number and collocation of their particles seemed to overcome that tendency and ended up producing the same shape. This remarkable regularity allowed chemists to use isomorphism as a way to ameliorate the conventionality of combinatorial units, and as a check on hypotheses about unknown compositions.[25]

Crystal surfaces have become important for reasons other than their arrangement. In particular, by the middle of the nineteenth century it was well established that metallic surfaces had the capacity to promote chemical combination, particularly if the surface-to-bulk ratio of metals like platinum was increased by fine subdivision or by using it in spongy form. The concept of *catalysis* was introduced to refer to this facilitating role and by 1850 it has become part of the consensus. It has also become common practice to use catalytic substances to bring about reactions that do not occur spontaneously. What remains mysterious is the mechanism of action of catalysis. The earliest proposal was that catalysts awakened dormant affinities. Another proposed mechanism was that substances undergoing decomposition could transmit their tendency to break apart to other substances with which they come into in contact. Finally, and closer to the real solution, were mechanisms involving the interaction of different attractive and repelling forces: adhesion to the surface of metals counteracts the repulsive force of heat that prevents gaseous particles attaining the proximity needed for the manifestation of affinity.[26]

Finally, the list of new concepts that are part of consensus practice by 1850 must include all those related to electrolysis itself: *electrolytes*, *electrodes*, *ions*. The terminology seems to be still in flux—the author uses the terms "platinode" and "zincode" to refer to negative and positive electrodes—but their referents are clear.[27] The concept of *polarity* has also become consensual. A polar force, such as the force exerted by a magnet upon iron particles, is one in which equal powers act in opposite directions. A solution of hydrochloric acid, for example, could become polarized by introducing into it a zinc wire connected to one end of a battery. Upon contact with the wire the components of the acid acquired a direction with the chlorine becoming negatively charged while the hydrogen became positively charged. If the circuit was closed through the introduction of a wire made of platinum, the resulting polarization became more extensive: hydrochloric acid decomposed into its elements and then recomposed again, forming a channel through which electricity could travel, the chlorine and hydrogen migrating in opposite directions. In order for this polar effect to take place, the dissolved substance (the electrolyte) had to be a compound since it was its decomposition into elements (ions) and their subsequent recomposition that performed the transmission of electricity. This implied that the electrical current partly neutralized affinity in one direction (causing decomposition), while it strengthened it in the other direction. It was this close connection with affinity that led some nineteenth-century philosophers to predict that polarity would one day become the great unifying concept.[28]

A variety of general statements describing chemical regularities in the action of these physical forces (heat, cohesion, adhesion, electricity) has become part of the consensus at this time. The list of statements includes the following:

1 When two substances react directly, heat is emitted. The faster the reaction the greater the amount of heat.

2 Adhesion between two dissimilar liquids causes them to mix by diffusion. The rate at which diffusion takes place varies in a regular way in different substances.

3 The amount of chemical action in electrolysis, measured by the quantities of decomposed substances, indicates the quantity of electrical power in circulation.

4 A circuit may be created with two different metals and a liquid. The metal with greater affinity for oxygen becomes the positive pole, and the one with the least affinity the negative one.

5 A circuit may be created with two electrodes made of the same metal if two different liquids are used that are in contact with each other and have different affinities for the metal.[29]

A large number of quantitative statements declaring the results of the measurement of various properties have also entered the consensus at this time. The referent of the 61 elementary substance concepts, and the much larger number of concepts for the compounds of these, is still fixed instrumentally, by specifying the separation and purification procedures needed to produce the substance in question. But physical and chemical properties are also used for their identification at different stages of the purification process. As in 1800, quantitative statements about these properties were not typically expressed linguistically but as *tables of values*. While these tabulated results can be considered to be a mere graphic representation of the corresponding statements, it can be argued that they also functioned as taxonomic devices, since they not only reported results but exhibited them in a form that aimed at revealing regularities. Moreover, even when the measured properties were purely physical, the *focus on variation* was characteristic of a chemist. In other words, whereas a physicist may be content with studying the relations between the properties of a single substance—using water or steam as representative of all liquids and gases, for example—a chemist had to pay attention to the differences between the properties of all elementary and compound substances, differences neatly exhibited in tabular form.

The main difference between the two textbooks in this respect is the sheer number of tables, and the number of entries per table: several more properties are being measured in 1850 for many more substances. The list of properties tabulated 50 years earlier included: saturation; dilation and expansion by heat; heat conductivity; heat capacity; latent heat; melting and boiling points.[30] Tables for these properties are still included in the new textbook, proof of the continuity between consensus practices, but new ones have been added: the optical refractive power of solids and gases; electrical insulating and conducting capacity; specific induction; electropositive or electronegative strength; specific electricity; magnetic power.[31] And the tables given for properties that were already part of the consensus have increased in the number of entries: the table for latent heats, for example, had nine entries in 1800 but 22 in 1850; the table for the expansion of solids by heat has roughly the same number of entries, but the one for gases has gone from seven to nine, and for liquids from six to 26; heat conductivity had been determined for only five solid substances in the previous textbook, while ten values are given in the new one.[32]

Electrolysis created many new phenomena that posed problems to chemists. First of all, the electrolytes themselves posed the problem of their susceptibility to decomposition by an electric current. Why some dissolved substance could play this role but not others was something in need of explanation. Known electrolytes tended to be binary compounds of non-metallic elements and metals, or metallic oxides, or even oxygen-based acids. This suggested an explanation in terms of polarities: binary substances have one electropositive and one electronegative component, and are therefore decomposed by attraction to the differently charged poles of a circuit. But several binary substances, like water or chloride of sulphur, were known to be incapable of acting as electrolytes, while yet others were made of more than two components. Hence, an explanation in terms of polarity did not offer a guide to understand their electrolytic behavior. So the problem at this time was: Why do some substances easily decompose by the action of voltaic current while others resist decomposition?[33]

The reasoning patterns used by chemists to think about problems of transformation have improved thanks to the differentiation of the three forces of attraction. As usual, we can reconstruct the relevant changes by using explanatory schemas. Statement A-4' of the Affinity schema is the most affected by this, since the mechanism of action of the counteracting factors can be better specified. Cohesion can counteract affinity forces by preventing the necessary intimate contact between the particles of two solid substances, and heat can do the same for gaseous ones. But a certain amount of heat (causing melting) or a certain degree of adhesion (of dissolved solid particles and those of the liquid solvent) can decrease the effects of cohesion and promote chemical union. In addition, after decades of neglect, the status of Berthollet's hypothesis that the degree of concentration of a substance can alter (or even reverse) the order of affinities has changed, the textbook confirming evidence from several quantitative studies.[34] Reasoning patterns about compositional problems, on the other hand, still show the tension created by the lack of instrumental means to separate mixtures from compounds. The condition of oxygen, nitrogen, carbon dioxide, and water in the atmosphere, as Dalton suggested, is conceived as a mere mechanical mixture. So is the combination of silicates of potash, soda, lime, magnesia, and alumina in ordinary glass, or the combination of different salts in mineral waters. But the status of metallic alloys is still problematic, some considered compounds with properties different from those of their components, while others are classified as mixtures with a blend of properties.[35]

Finally, there is a new cognitive tool imported from the outside: mathematical models. As we saw in the previous section of this chapter, the use of mathematics to model chemical reactions became common in physical chemistry as a new conception of heat as energy replaced that of an imponderable substance (caloric). By 1850 it is acknowledged that the treatment of other incorporeals, like electricity and magnetism, could benefit from the use of mathematical models. For simple electric circuits, dependencies between the properties involved—like the inverse relation between the conducting power of a wire and its length—could still be handled informally. But for compound circuits in which the plates were arranged in separate compartments and connected alternately, a rigorous treatment of the mutual action between electro-motive forces and the resistances offered by the wires, plates, and liquid solutions demanded the use of mathematics.[36] Thus began a slow infiltration of chemistry by equations developed by physicists, becoming a full-scale invasion in the last decades of the century.

1850–1900

Robert Wilhelm Bunsen (1811–99), Gustav Roger Kirchhoff (1824–87), Dimitri Mendelev (1834–1907), Wilhelm Hittorf (1824–1914), Jacobus van't Hoff (1852–1911), Svante Arrhenius (1859–1927), Friedrich Michael Ostwald (1853–1932)

The phenomena that constituted the chemical domain during the nineteenth century could be, for the most part, explained in macroscopic terms. Some domain items, like the phenomenon of isomerism discussed in Chapter 2, suggested that spatial arrangement should be added as a causal factor in the explanation of the properties of a substance, and the most obvious candidate for that which was arranged in space was atoms. Nevertheless, the connection was indirect. But two new phenomena that were added in this century created a more direct link to the microscopic world: *spectral lines and radiation*. The history of the former is linked to the use of prisms to analyze white light into its component colors which had been part of the practice of physicists since the seventeenth century. In the early 1800s it was discovered that this analysis could yield much more structure than just the rainbow colors: it displayed lines of different widths and sharpness, having well defined positions relative to the color spectrum.[37] By 1850, the existence of the phenomenon was accepted by the chemical community, although it was also agreed that there was no adequate explanation for it.[38] Chemists were familiar with *flame color* as a physical property that was useful to determine chemical identity: sulphur burned in oxygen with a pink flame; cyanogen with a purple flame; carbonic oxide with a blue flame; soda compounds with a yellow flame. But the novel structure added to the color spectrum promised to greatly increase the analytical power of this approach, as correlations began to be found between spectral lines and chemical composition.

Although primitive spectroscopes existed before 1850, using slits to focus the light and prisms to disperse it, the first instrument to contain all the components of a modern spectroscope appeared in 1859, a result of the collaboration of a chemist, Robert Bunsen, and a physicist, Gustav Kirchhoff. In 1855, Bunsen perfected a gas burner that provided a clean flame of very high temperature, making both emission and absorption effects easier to observe. With Kirchhoff he studied metallic salts as they burned, and discovered that if a light source of sufficient intensity was placed behind the colored flame, dark lines appeared in its spectrum, as if selected wavelengths of the light were being absorbed by the flame. By analyzing many compounds of metallic substances like sodium and potassium, they found that the characteristic lines for the metals were always there, implying that a particular arrangement of lines could be used to identify the presence of a metallic element independently of the other elements with which it was combined.[39] Several new elementary substances were identified in the 1860s using their spectral signatures, cesium, rubidium, thallium, and helium, many years before they could be separated and purified in the laboratory. In that same decade the first international congress of chemistry reformed the combinatorial units used in formulas, arriving at a more coherent set of atomic weights. The new conventions, and the more complete list of elementary substances made possible by spectroscopy and other techniques, intensified the *search for patterns* that could help bring order to this growing portion of the domain.

Six different practitioners joined the search, each one discovering on his own that the pattern in question was a *periodic repetition of the properties* of substances.[40] As we saw above, these efforts had been preceded by the discovery of triads of elements related by their weight values, an approach that was limited by the inaccuracies still plaguing the determination of those values. But as the search for pattern continued, it slowly dawned on some researchers that these inaccurate values could be used to *order the elements in a series* of increasing weight. Ignoring absolute weights and keeping only their serial order was a crucial insight allowing the grouping of elements by their position in the series in one direction, and by similarity of properties in a second direction.[41] On the other hand, there were several conceptual difficulties to be solved before this two-dimensional arrangement could be made final. One of them was that some elements possessed different properties in purified form and as a part of a compound, so the ordering along the second direction raised the question of what properties to use: should chlorine, for example, be classified as a green and poisonous gas, the form in which it exists as a purified substance, or by the properties that it exhibits when it is part of a compound, like salt? The Russian chemist Dimitri Mendelev decided to use the latter, because the properties that chlorine exhibits as part of a compound reflect better its kinship with other halogens, and the same is true for other traditional groupings.[42] Mendelev found the coherence of the two-dimensional arrangement so compelling that he dared to leave empty places for hypothetical elements which, if the taxonomic schema was correct, *should* exist. These gaps were, in effect, predictions that the missing elements would one day be found. In the original table of 1869 there were empty spaces for three elements, gallium, germanium, and scandium, all of which were subsequently isolated.[43]

Electrolysis was the main cause of the proliferation of elementary substances, but it was also capable of generating problems if treated not as an analytical instrument but as a phenomenon in its own right. This was particularly true for the relation between the solvent and the diluted electrolyte. The problem posed by this relation could only be correctly posed, however, after performing extensive measurements of physical and chemical properties and finding regular dependencies among them. An important property to be determined was the velocity with which the dissolved electrolyte, once broken into ions, migrated to the electrodes.

The oldest hypothesis was that differently charged ions would migrate to the poles at the same speed. This could be tested by linking speed to final concentrations: if ions moved with different speeds, the faster ones would gather at one electrode in greater proportion than the slower ones. In 1854, the German physicist Wilhelm Hittorf tackled this problem with an improved version of an existing method that used porous membranes to isolate the areas around the poles, facilitating the measurement of ion concentration. After studying many different electrolytes, he established not only that migration rates were not equal, but that the proportion of ions gathered at each electrode was independent of the intensity of the current, depending only on the initial concentration of electrolyte.[44]

To express the problems posed by quantitative dependencies like these, a new generation of chemists began to draw on mathematical tools, an instrument useful not only to describe known dependencies between properties, but also to discover new ones. The conceptual and practical prerequisites for a mathematical treatment of the relationship between solvent and solute were in place by 1873, but were dispersed throughout the chemical community: a concept of equilibrium as a dynamic balance between reversible chemical reactions; ways to measure the properties of actual reactions in this equilibrium state without disrupting it; and an equation to capture the characteristics of the equilibrium state. It took the work of Jacobus van't Hoff to bring these different strands together and weave them into a coherent research program.[45] Van't Hoff created both components of the mathematical model: the equations and the ideal phenomenon. We discussed van't Hoff's work earlier in this book, and in particular the way in which he was able to imagine the affinities of carbon arranged with a tetrahedral symmetry. His imagination was also on display when inventing chemical analogs of ideal phenomena, not so much picturing a liquid solution at the limit of low concentration as the chemical version of the ideal gas, as imagining cyclic phenomena to be the counterparts of the ideal engine: an ideally selective membrane dividing a container into two sections with different chemical potential, and a reversible chemical reaction taking place at the interface. He also showed how to derive the chemical analog of the microscopic model of dynamic equilibrium in 1884, using the concentrations of substances in a reaction, the number of molecules taking part in it, and the velocities of forward and reverse processes as its variables.[46]

Hittorf's studies of the rates of ionic migration had used very low electrical currents, suggesting that the dissolved substances needed only a small push to generate the sequence of chemical dissociations and reassociations that performed the chemical work in electrolysis. This, in turn, suggested that a portion of the electrolyte had become dissociated by its very contact with the solvent, existing in this dissociated form as electrically charged ions even in the absence of any externally applied current. To chemists this idea seemed counterintuitive. Nevertheless, the Swedish chemist Svante Arrhenius took the possibility seriously and began investigating it, publishing his results in 1884. At first he refrained from asserting the independent existence of ions in solution, and made the less controversial statement that the dissolved electrolyte was divided into active and inactive portions. He did, however, obtain experimental evidence that the lower the concentration of the electrolyte, that is, the *more dilute* the solution, the higher the proportion of particles existing in the active state, as indicated by their increased electrical conductivity.[47] This created a connection with van't Hoff's work, since the latter had used the idealization of extreme dilute solutions. It did not take long for Arrhenius to realize that his activity coefficient (the ratio between the number of actual active particles and the total number of possible active particles) was related to an arbitrary correction factor that van't Hoff had been obliged to add to one of his models, showing that the latter was not so arbitrary after all.[48]

The personal practices of van't Hoff and Arrhenius did not have an immediate impact on the chemical community, for several reasons: the community's focus on problems of organic composition and synthesis; its unfamiliarity with the mathematical approach; and van't Hoff's and Arrhenius's geographical isolation from the centers of research in which the community congregated. So another inhabitant of the Scandinavian periphery, Friedrich Wilhelm Ostwald, was the first to realize the true significance of their work. Early on in his career, Ostwald had set out to redirect chemical research away from a static taxonomy of substances to a dynamic study of chemical reactions.[49] He knew of the work of other practitioners who had performed accurate measurements of reactions in which different acids, when neutralizing a given alkali, evolved different amounts of heat. He extended this line of research to other properties of acids that changed regularly when they interacted with alkalis (volume, refractive index, reaction velocity), searching for correlations that could throw light on the nature of affinity.[50] He also knew that the regularities that he and others had discovered could only be linked to one another through the use of mathematics.

Arrhenius's claim that chemical activity and electrical conductivity were related suggested to Ostwald that the strength of an acid could be related to the degree to which the acid's particles became dissociated in solution: the particles of a strong acid, like sulphuric acid, should be more ionized than those of a weaker one, like acetic acid. To test this hypothesis, he measured the conductivities of many acids, correlating the results with other properties he had already measured, and adapting mathematical models from physics to search for dependencies among the properties.[51] This line of research made him an ally of the two other pioneers, the three together launching the first official journal of physical chemistry in 1887. Unlike his partners, however, Ostwald was a strong opponent of atomism. To him, modeling micro-mechanisms was entirely unnecessary, the macroscopic study of *energetics* being enough to satisfy the needs of both physics and chemistry.[52] Their partnership is one more example of a fruitful collaboration between chemists that was not made impossible by the incompatibility of their ontological stances.

Let's examine now a reference textbook to see what components of these personal practices had become part of consensus practice by the year 1900.[53] The most striking conceptual change is the reorganization of many old ideas around the concept of *energy*, defined as anything that can be produced from work or that can be transformed into work.[54] The properties of substances are now conceived as depending both on their material and energetic components. The textbook acknowledges that composition (nature and proportion of components), manner of union (as determined by valence), and spatial arrangement (proximity of radicals to each other) are determinant of a substance's properties.[55] But to these factors the energy content of a substance is now added as an additional cause. As argued in the previous section of this chapter, there was a precedent for this: the idea that the different properties of one and the same substance in different phases (gas, liquid, solid) depended on the amount of "caloric" in it. By 1850, heat had ceased to be viewed as an imponderable substance, as its similarities with light and electricity had become evident.[56] But in 1900, heat is not only explicitly defined as a form of energy, but the role of energy content in defining the properties of substances has become well established: energy content is used to differentiate polymorphs (substances with analogous composition and similar crystal shape); isomers (substances with the same composition but different properties); and substances that have been dissociated into ions to different degrees.[57] The very distinction between compounds and mixtures that had been so problematic since the beginning of the century can now be formulated in energy terms: if the interaction between two substances does not cause energy changes, they form a mixture; if it does, they form a compound.[58]

The properties of different energy forms have become more clearly characterized. The consensus of 1800 already included the idea that a substance can contain a greater quantity of heat than its temperature indicated.[59] This suggested a distinction between the quantity of heat and its intensity. Now, this distinction has been clarified and formalized: the properties of all forms of energy are classified under two categories: *extensive and intensive*. Thus, thermal energy has quantity of entropy as its extensive property and degree of temperature as its intensive one; electrical energy has quantity of current as its extensive property and degree of voltage as its intensive one; volume energy has quantity of space as its extensive property and degree of pressure as its intensive one. The difference between extensive and intensive properties is their behavior under the operation of addition: given two substances at the same temperature and pressure, each possessing an equal amount of entropy or matter, bringing them together doubles the amount of the extensive properties but leaves the intensive ones unchanged. Intensive properties are different when they behave differently: two substances at the same pressure or temperature have no physical effect on one another if brought into contact; but the same substances at different pressure or temperature form a *gradient* when in contact with each other and are capable of driving a flow of energy from the point of high intensity to that of low intensity.[60] Given this distinction, affinity can be conceived as an intensive property: differences in affinity can perform the work necessary to drive a flow of matter. Affinity can then be defined as the maximum of external work that can be performed by a chemical process, a quantity measured by the change of free energy.[61]

The concept of a state of *equilibrium* was explicitly accepted in 1800, and implicitly before that, because it entered into the explanation of the workings of certain instruments: mechanical equilibrium played this role for weight-measuring devices like the balance, while thermal equilibrium did the same for thermometers.[62] The state of equilibrium was conceived as the final product of a process, but the state itself was viewed as one in which no further changes took place.[63] By 1900, the concept has changed to that of a dynamic equilibrium, maintained by processes occurring in both directions at the same rate.[64] The fact that the stability of this steady state is dynamic implies that the state can be spontaneously arrived at, but it cannot be spontaneously left. Work must be performed to push a process away from equilibrium, by changing the volume of the reactants, for example, but then temperature and pressure will spontaneously find new values compatible with a new equilibrium. state.[65] In a sense, the concepts of equilibrium and stability have begun to diverge, a divergence not as noticeable in 1900 when the only type of stability considered is the steady-state type, but much more visible later on when phenomena exhibiting periodic (and even chaotic) forms of stability were discovered. But even at this point, the concept of stability has become richer, including various forms of *metastability*, as when a liquid substance is cooled past its freezing point but prevented from becoming solid by careful removal of any impurity that can act as a nucleation center for the formation of crystals.[66]

The concept of energy brought about these wide-ranging changes in conjunction with two basic thermodynamic statements (statements 1 and 2 below) which have become consensus in the chemical community. Statements 3 to 5 are chemical versions of the same statements using the concept of free energy:

1   The total energy of the universe remains constant.

2   The total entropy of the universe remains constant or increases.

3   The change in free energy of a chemical process depends only on its initial and final states, not on the path between.

4   The change in free energy of a chemical process is equal to the free energy used to unite the products into a compound minus that used to decompose the reactants.[67]

5   The free energy of a chemical process either remains constant or decreases.[68]

A set of related statements, known as *extremum principles*, has also been imported from physics. They all involve the idea that a spontaneous process is characterized by the fact that some property tends to either a minimum or maximum value. Some of these statements are just different versions (or consequences) of statements 1 and 2 above. It is important to show them in this form, however, because the further development of the thermodynamic approach (its extension to non-equilibrium states, for example) involved the discovery of further extremum principles:

6   A process is at equilibrium at constant temperature when its free energy is at a minimum.[69]

7   Any process that can be conducted reversibly yields a maximum of work.[70]

8   If a liquid is bounded by a gas it tends to reduce its surface to a minimum (forming a droplet).

9   If a liquid is bounded by a solid it tends to increase its surface to a maximum (wetting the solid).[71]

Testing statements like these in the laboratory could be done only once all energetic changes had taken place, that is, when all the gradients had been cancelled, and there was no further cause for the occurrence of an event. This implies that *time* was not taken into account, or that it was assumed to have no effect.[72] As we argued in the previous section, time was reintroduced into thermodynamics when far from equilibrium states were finally modeled. But there is one aspect of chemical reactions in which the temporal dimension cannot be neglected: the *velocity* with which a transformation takes place. A variety of statements have been accepted by 1900 regarding the velocity of chemical reactions:

10 Reaction velocity is a function of substance concentration and time. Because the substances in a reaction are constantly appearing and disappearing, velocity changes with time.[73]

11 Reaction velocity is proportional to the concentration of the reacting substance.[74]

12 Reactions begin with the greatest velocity and end with their velocity approaching zero, if temperature and pressure stay constant. If the reaction produces heat, then it begins slowly and then accelerates.[75]

13 Reaction velocity may be affected by small amounts of special substances (catalysts) that are either not transformed by the reaction, or reconstitute themselves at the end.[76]

The referent of substance concepts is still fixed instrumentally, like it has been for the last 200 years, by separation and purification methods, but properties and dispositions are also used for identification. A variety of statements describing the identifying characteristics of new elements and compounds has been added, but the manner in which these descriptions are arranged has changed. In the previous textbook, elements are still subdivided into two main groups, metallic and non-metallic, and the order in which non-metallic substances are discussed is dictated by their relative abundance in the planet (oxygen, nitrogen, hydrogen, and carbon) or possession of certain common properties, like combustibility (sulphur, phosphorous). The only elements grouped in a similar way to 1900 are the halogens: chlorine, bromine, iodine, and fluorine.[77] In our current reference textbook, on the other hand, the statements follow the regularities displayed by the Periodic Table. Its columns provide the order in which elementary substances are presented—eliminating the blanket distinction between metals and non-metals—as well as the order of description of their compounds: the compounds of halogens (halides) are described first; then the compounds of oxygen and sulphur (oxides, sulfides); those of carbon and silicon (carbides, silicides) after that, and so on. What seems clear is that the new taxonomic schema has increased the degree of order not only of the part of the domain constituted by elementary substances, but also the portion constituted by inorganic compounds.[78]

Electrolytic reactions continued posing problems, just as they had done throughout the century. The mathematical model that successfully described the behavior of some electrolytes, for example, turned out to have exceptions. The model asserted that electrical conductivity was an exact measure of degree of dissociation, but this statement was false for *strong electrolytes*, acids and bases that become completely dissociated in solution.[79] Thus, the following problems remained unsolved in 1900:

Why do strong electrolytes not exhibit an exact relation between electrical conductivity and degree of dissociation?[80]
Why does the degree of dissociation of a substance depend on its chemical nature?[81]

Other problems were posed by new information that has become available on the velocity of chemical reactions. The statements about reaction rates listed above are reports of laboratory results, but they still lack a full causal explanation. In some cases a causal factor is known—that temperature affects reaction velocity, for example—but the effect of others factors remains unknown. Thus, the following unsolved problems about velocities are recognized:

Why do different reactions proceed at different velocities?[82]
Why is the effect of catalysts on reaction velocity so specific?[83]
Why are catalysts that decelerate a reaction much rarer than those that accelerate them?[84]

The most novel addition to the cognitive content of the field is, of course, mathematical models. Among the practical reasons given for this inclusion is this: the velocity of a reaction varies as the concentration of reactants and products changes, so laboratory measurements may yield different results depending on the *duration* of the measuring operation. Hence, ideally, the measurement should be conducted at extremely short durations, and when this is not technically possible, mathematics should be used to calculate those values indirectly. In particular, one operator from the differential calculus (differentiation) can be used to calculate the instantaneous rate of change of a reaction.[85] Thus, at this point, the manipulation of symbols on paper to learn about processes has become accepted, joining the ranks of Berzelian formulas and other paper tools. In addition, while in the reference textbook from 1850 only a single equation appeared, in 1900 equations are used throughout the text to express numerical relations between measurements, and to study regular dependencies among the properties measured.

The final cognitive tool we will examine is phase diagrams. In our reference textbook, phase diagrams have become so important that they provide the backbone for the book.[86] As we saw, phase diagrams depict a space of possible combinations of intensive properties, each point in the diagram representing a combination of temperature and pressure, or of temperature and concentration. The use of diagrams to understand the outcome of chemical reactions is done by analogy: the state in which the final products of a reaction constitute a pure compound, and the state in which the products and reactants are mixed together, are treated as being analogous to different phases.[87] And much as those combinations that represent a dynamic equilibrium between two phases (gas and liquid, liquid and solid) form *limiting curves* in a diagram, so do those in which compounds and mixtures coexist.[88] The set of limiting curves, with their intersections and alignments, constitute the structure of the possibility space, a structure that can have explanatory value. Thus, if the region of the possibility space containing those combinations of temperature and concentration that yield pure compounds shades into the region with values that lead to the production of mixtures, the absence of a sharp boundary is used to explain why there are so few chemical reactions that produce perfectly pure compounds.[89]

4 SOCIAL CHEMISTRY

Conventions, Boundaries, and Authority

In the previous three chapters we have treated scientific fields as if their identity was defined by a domain of phenomena and a community of practitioners, connected to each other by laboratory procedures and instrumentation. But the historical identity of a field is also affected by a variety of social factors: conventions, promotional literature, authority effects, political alliances. The objective phenomena that compose a domain, moreover, need not always be created in private or academic laboratories but can be produced in industrial ones, a context confronting practitioners with a further set of social factors, from patents and trade secrets to economies of scale and the demand and supply of materials and finished products. These other determinants of a field's identity must be carefully articulated with the previous ones. In particular, we must avoid postulating the existence of *relations of interiority* between the different components, relations that constitute the very identity of their terms, leading to a conception of a scientific field and its social context as an organic whole in which technical and social factors are *inextricably related*. This holistic maneuver, fusing all components into a seamless totality, is what allows sociological approaches to ignore the cognitive content of a field, since this content is not conceived as possessing an autonomous existence. If instead we articulate the different components using extrinsic relations, relations that preserve the relative autonomy of what they relate, we must take cognitive content seriously, and focus on the local interactions between individual cognitive tools and individual rhetorical claims, genealogical myths, demarcation criteria, and other social factors.[1]

Let's begin the incorporation of social components into our model by analyzing the role played by social conventions. As we will argue below, some conventions (such as the rules of chess) are constitutive of the identity of the social activity they govern. But even in this case, they do not fully determine this identity, and interact in exteriority with other components, such as the tactics and strategy of chess players. Thus, it is important to carefully analyze the kinds of conventions that exist, and to avoid taking as our point of departure the unwarranted belief that social conventions imply a holistic approach, a mistake that has misled many thinkers in the past, constructivists and positivists alike.[2] The common holistic strategy of these thinkers is usually deployed as a solution to the problem of the *underdetermination* of theory by observation, or to be more exact, to the problem of choosing between rival theories when the available evidence does not compel a choice.[3] To set the stage for this discussion let's first describe what may be called the *standard model* of a theory. This model assumes that a theory is a homogenous collection of general statements, each one defining the set of conditions under which a phenomenon can be produced, statements like: If condition A, and condition B, and condition C obtain then the phenomenon X will always be observed.[4]

Evidence is modeled as a set of descriptive statements reporting the observation of phenomena in a laboratory. The relation between the two sets of statements is specified as being *deductive*: the statements comprising the theory deductively imply the statements reporting the evidence. Ever since Aristotle created his syllogism, we have been in possession of an algorithm to perform deduction, that is, to move truth or falsehood from general to particular statements. Because an algorithm is an *infallible* mechanical recipe, the particular statements derived from the general ones cannot be doubted; in other words, they compel assent. However, positivist philosophers have argued that given two competing theories entailing the same consequences, evidence alone is impotent to force a choice between them. One way to understand how such rivals can coexist in practice is through the concept of an auxiliary statement. Let's imagine a situation in which the theoretical and observational statements are something like this:

If heat is applied to a metal then the phenomenon of expansion will always be observed.
Heat was applied to a sample of this substance and no expansion was observed.

Given that the observational statement contradicts the theoretical one, it could be assumed that we are compelled to consider the latter falsified. But that is not necessarily correct. To move truth from the first to the second statement we need an auxiliary one like "The substance to which heat was applied was metallic." Faced with contrary evidence, a scientist may give up the original statement and conclude that expansion in volume does not always take place in metals, or the scientist might decide to abandon the auxiliary statement and conclude that the sample was not after all a sample of a metallic substance. The choice between the two alternatives, however, may be underdetermined by the available evidence, in which case, positivist philosophers would argue, the choice is arbitrary, guided by mere convention.[5] This conclusion is embraced by constructivist thinkers who add an extra twist: if the acceptance of a theory is not compelled by the evidence, if the choice between rivals is purely conventional, then only sociology can explain it. Their slogan can be expressed as "If it is not logical then it is sociological."[6]

We could accept that there are situations in which the evidence leaves a choice open between a small number of plausible hypotheses, but constructivists and positivists alike go on to make the much stronger claim that *all hypotheses* are reconcilable with the same evidence.[7] To put this differently, we could accept local underdetermination, but not the global underdetermination peddled by skeptics of either camp. There are various ways—some logical, some sociological—in which global underdetermination may be defended. A positivist thinker can devise a way of generating an infinite number of alternatives by demanding only that the alternatives be *logically possible*. Thus, in the example above, statements like "The substance to which heat was applied was a metal from another dimension" or "The type of heat that was applied possessed other worldly properties" would be considered plausible alternatives. But clearly, no scientist would ever take these auxiliary statements seriously.

Constructivist thinkers take a different route. In one approach it is argued that the standard model presupposes the *stability of the meaning* of the words composing the theoretical and observational statements. In other words, in the standard model the two sets of statements are treated as if they existed in a disembodied state with their meanings unchanged by time. But meanings do change, and the way in which they change can be underdetermined by past usage. The meaning of terms like "metal" and "heat," for example, changed in the past: the first term used to mean a compound substance before meaning an elementary substance, while the second used to mean an imponderable substance before meaning a form of energy. But if the very meaning of the words used in theoretical statements changes, does not that change the set of evidential statements entailed? And if so, does not this prove that, given appropriate changes in meaning, any set of theoretical statements can be made to imply any set of observational statements?[8]

The positivist line of argument can be blocked by arguing that what is logically possible can be, and often is, entirely implausible given the lessons of past practice. Even if the alternatives were not as outrageously implausible as the ones given above, statements offered as candidates to save a particular theory from contrary evidence may bear *cognitive costs*. For example, if the proposed auxiliary statements replaced others that occurred as part of explanatory schemas, like the Part-to-Whole or the Affinity schemas, and if the replacement lowered the explanatory power of the schema, this would count as a cognitive cost that the proponent of the modified theory would have to bear. Thus, the logical approach to generating an infinite number of alternatives relies for its plausibility on ignoring the costs of change.[9] The constructivist line of argument, in turn, can be blocked by arguing that the *meaning of a concept does not determine its referent*. In the case of statements asserting the existence of a causal relation between applied heat and the expansion in volume of a metallic substance, the referent of the latter is fixed instrumentally, by the separation and purification techniques used to produce the substance. This is why we can confidently assert that when chemists used the term "vitriol" in 1700, while later ones used the term "sulphuric acid," they were both referring to the same substance, even though the meaning of the two terms is entirely different. A similar point can be made about the referent of the term "heat": it was fixed by the variety of causal effects this form of energy could have, effects that all chemists were familiar with and could produce at will, even if they had the wrong hypotheses about their causes.[10]

We may conclude that the constructivist and positivist responses to the existence of underdetermination—the wholesale introduction of conventions—is unwarranted. The historical record does contain many examples of evidence failing to compel assent, but this only proves the existence of local, not global, underdetermination. A good illustration is that of the rivalry between the radical and type approaches to taxonomy in organic chemistry. Both theories had evidential support, although as we saw, the radical theory had been challenged by a piece of negative evidence: chemical reactions could be performed that substituted an electronegative element (chlorine) by an electropositive one (hydrogen), contradicting a statement in the theory. Nevertheless, both theories were considered serious alternatives for a long time. But while two or three plausible candidates may coexist for a while, this underdetermination tends to be transitory because evidential support is *not limited to logical consequences*. New evidence may be produced that is not logically related to any theoretical or observation statement being debated. Thus, at the end of the nineteenth century the debate between atomists and anti-atomists remained unresolved by the chemical evidence that each theory entailed. But in time even ardent anti-atomists like Ostwald were convinced, and were so by evidence that was produced while investigating a non-chemical phenomenon: Brownian motion. If new evidence can be provided by logically unrelated phenomena then it can be asserted that *the field of evidence is open*: a choice between two theories may not be determined at one point in time but it may later become fully settled by evidence from unpredictable sources.[11]

Much like underdetermination might be accepted as a local and transitory phenomenon, that arbitrary conventions play an important role in scientific practice can be admitted without agreeing with wholesale conventionalism. Let's begin by discussing the nature of conventions. First, for there to be a convention there must exist alternatives, but the degree to which the candidates are arbitrary is often constrained. Most conventions play a well-defined role in a social activity, and a *serious* alternative is one that could also play that role.[12] Thus, the conventions involved in the activities of purifying substances, measuring properties, or assessing capacities can have alternatives, but any serious candidate must be able to play the role of a standard of purity, a measurement unit, or a reference substance. Second, given a set of serious alternatives, it is not necessarily the case that all must be *equally valued*, although for there to be a convention no single alternative can be valued more than the fact that the convention is collectively used.[13] The different values attached to the alternatives, on the other hand, affect the degree to which a choice between them is arbitrary. A good example is temperature units: the choice between Celsius and Fahrenheit degrees, both of which use water's freezing and boiling points as standard reference, is conventional, and we know this because we can imagine using many other substances as a standard. But once the results of many measurements are plugged into a mathematical model the units begin playing another role and new reasons to prefer one alternative over the other emerge: using absolute temperature, in which the zero degree is set at -273 degrees Celsius, lets the ideal gas equation be simplified and exhibit in a more vivid way the relations among properties.[14] This shows that participants in a social activity need not be indifferent to all the alternatives, and hence that their choices can be guided by reasons.

Finally, it is important to emphasize that not all conventions work in the same way. We must distinguish between *constitutive* conventions that define the very identity of an activity, as exemplified by the rules of chess, from *coordinative* conventions that do not.[15] Changing a single rule of chess—moving a particular piece on the board in a novel way, for example—changes the identity of the game being played. But changing the units of measurement does not change the identity of the practice of measuring properties, except trivially.[16] Another difference is that while coordinative conventions tend to *regulate pre-existing practices*, and often do so using a single rule, constitutive conventions *create novel practices* through the use of a complex set of rules.[17] Moreover, as these novel practices spread, they tend to generate sets of values that increase their complexity as a social activity. Thus, although the rules of chess do not specify the standards used in the evaluation of players—what counts as an elegant move or as a brilliant strategy—the values guiding these assessments have been generated as part of the process of engaging in the rule-governed activity.[18] The question now is whether the culture of laboratories is like that of chess, fully constituted by its conventions and generating its own values, in which case the conclusion of constructivist historians that experimental practices are based on arbitrary foundations would be correct.[19]

There is no way to answer this question other than by examining the historical record *over periods of time long enough* to eliminate the artifacts created by short episodes of underdetermination. Episodes like these can occur when there is a single laboratory instrument that can produce a given phenomenon. In this case, the phenomenon will be defined as real if it is replicable using the instrument, while a correctly functioning instrument will be defined as one in which the phenomenon can be replicated. This circularity suggests that the very phenomenon is constituted by a convention.[20] But although episodes like this do occur in the history of chemistry, they tend to be transient, and the practitioners themselves seem fully aware of the unreliability of the results produced under those circumstances. Constitutive conventions do play a role in scientific practice, as we will argue below, but this role is limited. And just what effect constitutive conventions do have must be examined case by case, cognitive tool by cognitive tool, a requirement that cannot be met by constructivist methodologies that deliberately ignore a field's cognitive content.

The majority of conventions used by chemists in the period studied by this book are of the coordinative type. These conventions, as was just mentioned, regulate pre-exiting practices, so they do not enter into the constitution of their identity. The purification of substances, for example, was practiced by pharmacists, metallurgists, and alchemists long before chemistry existed as a field—and similarly for the practices of measuring properties (volumes) and assessing capacities (like the medical virtue of medicines). Even those standards born as part of chemical practice, such as the definition of an elementary substance as the limit of chemical analysis, were used in practice for decades before Lavoisier made them explicit in an effort to increase coordination among different chemists. Another interesting case is that of the combinatorial units (equivalents and atomic weights) used by nineteenth-century chemists to establish the relative proportions of the components of a substance. These two rival ways of performing the task evolved alongside the practice itself, never constituting its identity, and were relatively interconvertible like units of temperature. Moreover, every chemist involved understood that both units were conventional, and took that into account when assessing their relative merits.

To understand the role played by coordinative conventions, we need to define them in more detail. In an influential model, the kind of social activity regulated by these conventions is said to involve four different components: a set of personal actions; a desired outcome determined jointly by all the actions; personal choices based on expectations about other participants' choices; and finally, a set of alternative pairings of actions and outcomes, one of which is preferred not because it is valued more by all participants, but because no one can improve the collective outcome by unilaterally changing to a different alternative action.[21] In the case of combinatorial units, the personal actions were those involved in establishing relative proportions; the desired collective outcome was a situation in which results from different laboratories could be compared with one another; the choices were over rival combinatorial units, based on the expectation that others would be using the same unit; and finally, everyone involved had a choice between different ways of establishing proportions, all compatible with the desired outcome of comparability of results, but once they had made a choice leading to this outcome, no one had anything to gain by unilaterally switching to another standard. Also, and despite the fact that different chemists had reasons to prefer one or another of the alternative combinatorial units—atomists tended to prefer atomic weights, while anti-atomists leaned towards equivalents—very few chemists valued the intrinsic merits of a standard more than the possibility of successfully coordinating their practices.[22]

One way of achieving coordination in cases like these is to make *explicit agreements*. This actually happened in 1860, as part of the first international conference of chemistry, in which most participants agreed to use the newest version of atomic weights in their formulas. This made their personal practices more coherent with one another. But for several decades before that, the community had spontaneously coalesced around two standards, excluding other possibilities, without any explicit accord. How this could happen can be understood by analyzing the dynamics of situations in which people's choices depend on what they expect others to choose. As long as each participant can reasonably expect others to choose on the basis of the desired collective outcome, and as long as each participant can be reasonably sure that all the others know this, they have a reason to coalesce around one of the alternatives. In other words, as long as there exists a set of *concordant mutual expectations*, and the required common knowledge about these expectations, the coordination problem can be solved without an explicit agreement.[23] Furthermore, it can be argued that the role of the agreement of 1860 was not to force a standard on the community, but to provide everyone with the correct mutual expectations and common knowledge.[24]

This formal model shows that coordination problems can be solved spontaneously. But to apply it to an actual historical episode, we must bring other factors into play. Atomic weights dominated the 1830s, due to the meticulous analysis of over 2,000 substances conducted by Berzelius. But in the following decade many members of the community drifted towards the equivalents proposed by the German chemist Johann Gmelin. The cognitive values invoked in this switch were related to the different technical means to assess relative proportions—atomic weights involved volumetric techniques while equivalents involved gravimetric techniques—so any technical problems with one approach reduced its value and increased that of the other. But other cognitive values were also involved. The units were used as part of empirical formulas and it was desirable that the latter were consistent for different portions of the domain. When Gerhardt became aware that the use of equivalents gave inconsistent formulas for organic and inorganic compounds, he proposed a new version of atomic weights and reopened the debate, causing the community to split in the 1850s. A partial consensus was reached by explicit agreement in 1860, the Italian chemist Stanislao Cannizzaro playing a crucial role in articulating and making explicit the cognitive values involved in the choice. Later on, when Mendeleev used the new units to reveal deep regularities in the set of elementary substances, his Periodic Table became evidence that the reasons offered by Cannizzaro had been sound.[25] This shows that, much like general statements may receive confirmation from observation statements not logically entailed by them, so choices over conventions can receive support, and decrease their degree of arbitrariness, from unsuspected sources.

We can conclude from this brief discussion that the conventions used in scientific practice are typically of the coordinative type, and that their arbitrary nature can be limited by their interaction with cognitive tools and the values associated with them. But does this mean that the other type of conventions is never a component of a scientific field? A possible example of constitutive conventions are those that define ideal phenomena: the ideal gas, the ideal engine, the ideal reaction. In this case, the argument that meaning does not fix reference cannot be used, because ideal phenomena have no real referents.[26] But does this imply that they are entirely arbitrary? And if so, what does that say about the cognitive status of the statements made about them? To answer this question it will be instructive to get back to the example of the game of chess. Its rules are, indeed, constitutive, but they are not the only factor determining the identity of the practice of playing chess. The tactics and strategy of chess include a variety of board patterns that are an important part of the game: different opening moves; various endgame situations; gambits in which pieces are sacrificed for later advantage. These patterns are made possible by the rules, but are underdetermined by them.[27] And a similar point applies to ideal phenomena: while the ideal gas may be entirely constituted by its conventional definition, the sets of equations that can be built around it are not. The cognitive value of each equation lies in its ability to reveal patterns of dependencies in the way properties change, and once linked up into families of models they can lead to the discovery of unsuspected new dependencies. Moreover, patterns can be discovered in the families themselves, such as certain symmetries and invariances. Neither the way in which different ideal phenomena are related to each other, nor the overall regularities that they display when so related, is constituted by convention.

Before leaving this subject, we must discuss the special case of explanatory or classificatory alternatives in which the choice between them is *not* conventional. This situation can arise because most objective phenomena are characterized both by their properties as well as by their capacities. Unlike the former, which are typically finite in number and can be listed, the latter form a potentially open set because their exercise involves coupling a capacity to affect with a capacity to be affected, and the potential couplings are multiplied as the contents of a domain grow. Thus, when a new substance is synthesized, the existing substances with which it could react, and the products of such a reaction, cannot be listed in advance. And the substances not yet synthesized with which it could react form an open-ended set.[28] Because chemical reactions can be used to classify substances into series, as in the type or radical approaches to classification, the openness of the set has consequences for alternative taxonomies. As the French chemist Adolphe Wurtz argued in the 1860s, different reactions can lead to different formulas. Thus, when giving a formula for acetic acid recalling its formation from a reaction of acetyle chloride and water, the radical C_2H_3O should be used. But when the formula is based on the action of soda on cyanide of methyl, the radical must be written in a different way, as $CO(CH_3)$.[29] The alternatives in this case are not mutually exclusive but complementary, the multiplicity of formulas representing not a fact about us and our social conventions, but a fact about the components of a domain and their multiple capacities to affect and be affected by one another.

In addition to social conventions, we must include in our model of a scientific field *mythologized genealogies, promotional literature, and rhetorical demarcation criteria*. These other social components define the image that a community of practitioners has of itself, as well as the image of the field that these practitioners project to other communities. They are therefore important in defining the historical identity of a field, but to understand their contribution they must be conceived as related in exteriority to the other components already discussed.[30] To begin with, the historical identity of a field must be preserved over time by relations of *legitimate ancestry and affiliation*: scholastic genealogies involving masters and pupils; institutional genealogies of laboratories, professional societies, and university departments; and technical genealogies defining families of instruments and machines.[31] The interaction of teachers and students is clearly the most central, since it gives the cognitive content transmitted across generations its continuity and authority, but it takes place in an organizational context that changes over time, and in relation to a domain of phenomena accessed through concrete instrumentation. When the genealogy of a particular field is codified as *official history*, however, it often contains elements of mythology, as when chemists characterized the origins of their field in terms of the heroic battle of Lavoisier against the irrational defenders of phlogiston. This view of the controversy is, as we have already shown, deeply mythological, but the narrative of the battle between reason and unreason constituted an item that was passed from teachers to students alongside the cognitive content of the field. Snippets of this mythologized history were typically present in the speeches given in a variety of traditional events: from graduation ceremonies and the opening of conferences to retirement celebrations of the leaders of schools or founders of sub-disciplines.[32]

Practitioners also had to project a particular image of their field to outsiders in order to obtain *external recognition* from other communities and organizations.[33] The personal practice of several chemists involved the deliberate promotion of chemistry as a field, but the recognition they sought was from different target audiences. Practitioners needed, first of all, the recognition given by their government to scientific organizations, endowing the latter with institutional *legitimacy*. They also wanted the recognition of communities of industrialists or of instrument makers, in which case it was not a matter of seeking legitimate credentials but an acknowledgment of the field's practical *utility*. Finally, there was the recognition of a field by other scientific fields, in which case the social value at stake was not related to accreditation or appreciation, but to relative *prestige*. Thus, the search for external recognition called for a variety of strategies, a variety made even greater by differences in local conditions. In Germany and Sweden, for example, where chemistry was linked to mineralogy and had proved its utility in many assaying and testing offices, promoters felt the need to stress its value as a university subject. It was in Sweden that the distinction between pure and applied science was introduced to reassure academicians that the subject was distinguished enough for them to teach.[34] In France, chemistry emerged in close association with pharmacy, proving itself as a viable alternative to traditional ways of producing medicaments. Thus, the promotional literature in the early eighteenth century stressed its unique ability to use analysis (distillation) to learn about a substance's composition, and the application of this information to the creation of substances with superior medical virtues.[35] Later in that century, the variety of ways to promote chemistry increased. On one hand, there were chemists like Guillaume-François Venel who did not feel that the hard labor involved in lengthy experiments, or even the dangers of handling poisonous or explosive substances, was in need of any apology: a true chemist should be able to take off his academic gown and engage with the messy materiality of substances.[36] On the other hand there were chemists like Macquer, who used the language of natural philosophy to create a lofty public discourse for chemistry, seeking to increase its prestige without surrendering its autonomy.[37]

In addition to a mythological self-image and a promotional public image, the boundaries of a scientific field are affected by statements about methodology. Grouping questions about scientific method together with legends and propaganda may seem strange, but questions like these tend to occur in the context of *presenting and justifying* the cognitive content of a field, and in this context the discussions tend to be mixed with rhetoric.[38] Moreover, from the start, practitioners of different fields had to deal with a role for methodological statements they had inherited from the distant past: their role as *demarcation criteria*. In particular, it was common for early practitioners to accept the two criteria that had been proposed by Aristotle, one to distinguish knowledge from opinion, the other to distinguish knowledge from craft. Knowledge and opinion could be demarcated by the ability of the former to deliver apodictic certainty, while knowledge could be distinguished from the crafts by its ability to comprehend the ultimate causes of a phenomenon.[39] As with the other two components, views on method developed in Aristotle's shadow displayed a great deal of diversity.

Most natural philosophers accepted the first criterion but differed in the way in which they proposed to implement it: to some the use of mathematics ensured that conclusions to arguments had the compelling power of deductive logic, while for others inductive logic, when used to eliminate alternatives from a well-defined set, could also go beyond opinion, although the certainty it provided was not incontrovertible but only practical.[40] The second criterion did not have the same general acceptance, and attitudes towards it also varied. Natural philosophers believed that the job of mathematics was to describe a phenomenon, not to explain it, and gave up the search for causes. Others accepted the need for causal explanations, but rejected the idea that artisans were unable to provide them: a catalog of the way in which they *produced effects* was an important source of information for anyone searching for the causes of those effects.[41] Given these conflicting constraints, even the most eminent natural philosophers had to resort to rhetoric to satisfy them when making methodological statements.[42] Moreover, statements used to exclude some activities as non-scientific tend to be *meta-statements*, and while chemists have proved to be reliable producers of information about substances and reactions, they have never displayed the same reliability when making claims about the nature of their own field—and neither have natural philosophers or their physicist descendants. Hence, picking a methodological statement by a single practitioner, however famous, and declaring it to be a constitutive convention determining once and for all what "counts as science" is a gross mistake.[43]

Does the variability of these social factors imply that their effect on the identity of a field is minor when compared with the role played by a shared domain, a shared instrumentation, and a shared set of consensus cognitive tools? It is impossible to answer this question in general, but we can perform an analysis of some key concepts that have traditionally been used as part of foundational myths, promotional literature, and demarcation criteria to check how hard it is to separate their cognitive content from their rhetorical crust. In this exercise it will be useful to employ as a point of reference a particular social context in which the content produced by natural philosophers was used superficially and uncritically for political and religious agendas: *scientistic movements*. Ever since the seventeenth century, groups of non-practitioners have striven to make the values they associated with science acceptable to other communities and organizations.[44] Although the content of scientistic movements is also highly variable, they shared a belief in the truth of five statements: first, that all scientific fields constituted a single unified entity; second, that there were no limitations to what could be explained using science; third, that past scientific achievements were evidence of this unlimited power; fourth, that the objectivity of science was guaranteed by the use of a single method; and fifth, that the products of scientific practice were unambiguously beneficial.[45] On top of these core statements there were a variety of others giving each form of scientism its own characteristics.[46]

The discourse produced by scientistic movements neatly isolates the cognitive tools most affected by mythological, propagandistic, and rhetorical uses. From this point of view, no statement has had a greater impact on scientism than the claim that science is characterized by the use of *unbiased observation* to discover the *eternal and immutable laws of nature*. There is not a single scientistic thinker who did not pretend to have discovered "social laws," like the laws that govern the historical or intellectual development of humanity, the existence of which could be established through a dispassionate observation of history.[47] These two concepts, eternal laws and unbiased observation, will serve as our case study in our analysis of the relation between cognitive and non-cognitive components. The concept of natural law played a public relations role in the seventeenth century, as natural philosophers sought recognition from government and ecclesiastical authorities: they gained legitimacy in the eyes of these authorities by arguing that their goal was to discover the laws that God himself wrote in the book of nature, using the language of mathematics.[48] Similarly, they enhanced their self-image by thinking that their mathematical models captured timeless and unchanging truths, and that possession of such truths was a necessary condition to be a real science. Influential philosophers used this criterion to demarcate physics from other fields. Chemistry, in particular, was marked off from proper science and assigned the inferior status of a systematic art because it did not possess a deductive backbone based on mathematics.[49] The question now is whether this rhetoric affected the cognitive content of the concept of natural law.

There are two types of cognitive tool to which the term "law" is normally applied: statements expressing empirical generalizations, and mathematical generalizations deemed to be *exceptionless*. Empirical generalizations are derived through induction, while their mathematical counterparts are, as natural philosophers used to say, deduced from ideal phenomena.[50] The fact that one and the same term is used to refer to two very different cognitive tools is itself a sign that rhetoric is at work: it is an attempt to endow generalizations that do have exceptions—such as the laws of definite proportions and multiple proportions that justified the use of integer numbers in the formulas of Berzelius—with the same prestige possessed by mathematical laws.[51] But the rhetoric also works in the other direction: strictly speaking, mathematical laws are true only of ideal phenomena, so it can be argued that they are literally false when applied to natural phenomena.[52] Only in a laboratory, where artificially simplified phenomena are created, can it be said that the mathematical law fits reality. This limitation does not detract from the cognitive value of mathematical models, a value that consists in generating families of models exhibiting definite symmetries and invariances. These regular patterns allow practitioners to create a framework that unifies a variety of previously disjointed true statements, as well as to discover unknown dependencies in the way properties change.[53] But this is very different from the rhetorical claim that they capture eternal and immutable truths about nature, a claim that is like a *theological fossil* left over from the time when all natural philosophers were deeply Christian.

A similar analysis can be made of the other pillar of scientism: the idea of an unbiased observation. First of all, with the possible exception of astronomy, the role of observation in the production of evidence is relatively limited, compared to the precise measurement of the properties and capacities of phenomena. Some positivist thinkers believe that they can reduce everything to observation by arguing that the measurement of unobservable properties like temperature or pressure can be reduced to the observing of markings in a thermometer or a barometer.[54] But this description of the practice of precision measurement is a caricature. The precise determination of weight in chemistry, for example, depended on two factors: an increase in the sensitivity of balances, determined by the smallest differences in weight that the instruments could detect; and the creation of numerical methods to *extract information from the errors* that are inevitably made during weighing. These numerical methods, such as the method of Least Squares created at the turn of the nineteenth century in the field of astronomy, had been adopted by chemists by the 1840s as an aid in the determination of atomic weights.[55] The method begins by separating errors into those that are constant or systematic from those that seem to have no pattern and can therefore be considered accidental. The cognitive content of the term "unbiased" in the phrase "unbiased observation" refers to the removal of systematic biases by methods like these, not to the idea that the practice of scientists is dispassionate or disinterested, an idea that is entirely rhetorical.

One source of constant errors is imperfections in the instruments used to measure properties, a bias that can be eliminated by following certain protocols. When a chemist weighed the substances at the start and the end of a chemical reaction, for example, he might have noticed that the two arms of his balance were not of exactly the same length, and that this introduced a constant bias in the results. The protocol in this case was the procedure of *double weighing*, in which the weights on each side were reversed, and differences were estimated by adding small weights to the deficient side.[56] In the case of accidental errors, on the other hand, a similar approach cannot be used because their sources are multiple and unrelated. These characteristics, however, suggest that a *population* of such errors will tend to be randomly distributed, that is, that the variation in the measurements has a distinct statistical shape. Given this assumption, a recursive method using as its input an entire population of measurements can find *a collective value with an error that is smaller than any individual measurement*.[57] This procedure is the heart of Least Squares, a procedure that allowed practitioners to create statistical models of their data. Least Squares also provided them with the means to combine measurements performed at different laboratories under different conditions: different estimates for atomic weights that had been produced over a number of decades, for example, could be statistically combined in the 1880s to obtain more refined values.[58]

The fact that Least Squares possesses real cognitive value does not imply that the concept of precision measurement itself was devoid of rhetorical uses. Before chemists began learning from errors, for example, the increased precision of their instruments and procedures had already led to reports containing quantitative statements. But since the numerical expression of the results of chemical analysis was not accepted as an improvement by the whole community, rhetorical claims accompanied the numbers to make them persuasive. Lavoisier, for instance, reported the results of his analysis of water with eight decimal places, but his British counterparts quite correctly saw in these figures an implausible degree of exactness.[59] The method of Least Squares eventually helped to eliminate this kind of rhetoric through the concept of *margin of error*, a quantity that could be calculated for large data sets since 1816.[60] This quantity determined the maximum degree of accuracy with which the most probable value should be expressed, a degree that could not exceed the margin of error if the extra digits were to be considered *significant* figures.[61] A different source of rhetoric accompanied the meticulous protocols used to eliminate constant errors, because trusting their results implied trusting the people who used them.[62] Justifying the legitimacy of the numbers in this case went beyond mathematics and made use of existing rhetorical strategies common in other institutional contexts: a stress on the moral virtues of responsibility, due caution, and logical consistency.[63]

This brief analysis shows that it is possible to recover the cognitive content of even the most rhetorical scientific concepts and statements.

Removing the rhetorical crust, however, is fatal for scientism. We listed above the five statements accepted by most nineteenth-century scientistic thinkers, but we failed to add that analogs of all five of them can be found in the philosophical movement that gave us the standard model of theories in the twentieth century.[64] If this is correct it implies that the positivist and constructivist arguments over underdetermination, and the wholesale conventionalism they propose as a solution, are not arguments that bear on scientific practice but on scientistic thinking. Let's therefore strengthen the case for the scientistic character of the standard model. To begin with, the model depends crucially on the concepts of exceptionless laws and of disinterested observation. The first component of the model, the set of theoretical statements, must contain many general laws if it is to serve as the basis from which observation statements can be deduced. The fact that the term "natural law" refers to two different cognitive tools creates problems for this component of the model. If the term is taken to refer to empirical generalizations that have exceptions then the certainty associated with deduction is compromised. If, on the other hand, the term is used to refer to mathematical equations then the problem is that these *are not statements*. We can express the content of a given equation using language, but its linguistic expression lacks some of the cognitive features that makes the equation valuable: its capacity to be combined with other equations to generate new models; its capacity to be used as an input to mathematical procedures that reveal features of cognitive significance, like a distribution of singularities (maxima and minima); and finally, the invariants displayed by the equation under groups of transformations, invariants that explain why the equation can retain its validity under different circumstances.[65] The other component of the standard model, the set of observation statements used as evidence, depends on the idea of a detached and impartial observation to ensure its credibility. But what is compared with the predictions of a mathematical model are not reports of raw observations, however dispassionate, but *another model*: a statistical model of the distribution of measurement errors.[66]

We may conclude that neither constitutive conventions nor myth, rhetoric, and propaganda can distort the cognitive content of scientific fields, at least not to the extent that future generations cannot repair the damage. And this brings us to our final subject: the social factors that affect how this content is transmitted from teachers to students, and how the passage from personal to consensus practice is performed through this transmission. The vignettes used as illustrations of personal practice in previous chapters clearly display their variability, suggesting that the formation of consensus is performed via a *critical selection of variants*.

This selection, in turn, implies the existence of a variety of professional forums (academies, journals, meetings) supplying practitioners with the opportunity for critical debate.[67] Discussion in these forums can lead to the collective appraisal of variants but always in comparative terms: not what is the best way to conceptualize, describe, explain, or classify, but how a concept, statement, explanation, or taxonomy performs relative to its plausible rivals.[68] The critical appraisal of variants, however, can be constrained by *the effects of authority* in these forums. The editor of a journal, the president of an academy, the dominant voices in a meeting may exert undue influence on a discussion, prematurely fixing some variants at the expense of others. The extent to which this occurs can be correlated with the degree to which authority is centralized: the greater the degree of centralization the more likely it will be that the opinion of a few practitioners will prevail in a debate. The first step in a discussion of the role of authority, therefore, must be an examination of the factors that contribute to centralization and decentralization.

It will prove useful to have a simple model of the way in which the selection of one variant over another can be carried out in a critical way. This model of consensus formation has three components: the cognitive tool whose rightness of fit is being evaluated; the formal and informal rules of evidence that guide this evaluation; and the research goals that justify these rules as a means to an end. In this model, disagreement about rightness of fit—dissensus over whether a given concept has a referent, or whether a particular statement is true, or whether a specific problem is correctly posed—can be settled through arguments that invoke the rules of evidence; while disagreements about the rules can be settled through arguments that show that some evidence rules promote the achievement of a research goal better than others. There are two misconceptions that must be avoided to correctly formulate this model. The first is, as before, holism, conceiving of the three components as so inextricably related that dissensus about one component implies dissensus about the other two. This can also be sidestepped by treating the relations between the components as relations of exteriority, so that disagreement about one can coexist with agreement over the other two. The second is hierarchy, conceiving justification as flowing downwards from goals to rules, and from rules to concepts, statements, explanations, or classifications. This can be avoided by treating justification as also capable of flowing upwards: new phenomena or new cognitive tools may lead to re-evaluations of the rules of evidence, while new ways of gathering and assessing evidence may lead to re-evaluation of the extent to which a research goal, like the goal of attaining incontrovertible certainty, is in fact achievable.[69]

Let's illustrate this model with some concrete examples, keeping in mind that the collective agreement in question is never about an entire theory but about individual cognitive tools. For most of the eighteenth century, a shared research goal among chemists was the *discovery of the qualitative composition* of substances. This goal pre-existed the birth of the field, since the practice of pharmacists, metallurgists, and alchemists already aimed at determining a substance's composition in a qualitative way. Given this inherited goal, chemists developed informal rules of evidence to facilitate its achievement. The postulated composition of a compound substance, for example, could be determined by chemical analysis and *validated* by chemical synthesis. The justification for the analysis/synthesis rule of evidence was the argument that breaking apart a compound leads to the disappearance of its properties, and that a good test that the broken pieces are indeed its components is to put them back together, getting the original properties to re-emerge. If the substance in question happened to be elementary, the analysis/synthesis rule could not be applied, but a new rule could be created: if a substance resists all attempts at analysis by chemical means that is evidence that it is not a compound. These two rules of evidence often led to the achievement of consensus about the composition of many substances, but they were not enough to compel assent in all cases. In other words, the rules locally underdetermined choices. Thus, the inability to carry chemical analysis further could be explained by arguing that the substance was elementary, or by arguing that the analytical resources were insufficient in this case. Nevertheless, even in these underdetermined cases, the rules eliminated many alternatives, leaving only a few rivals to choose from.[70]

In the last quarter of that century, as the balance improved and the means to conduct closed chemical reactions in which nothing was lost were developed, the goal of *discovering the quantitative composition* of substances became feasible. The need for quantitative information was particularly important in the case of organic compounds because they all shared the same components in different proportions, but compounds of organic origin prevented the application of the previous evidence rule because chemists did not know how to synthesize them. Thus, the new research goal demanded new rules, like a strict balance sheet approach to the performance of experiments to ensure that no substance had entered or escaped. In addition, combinatorial units had to be devised to express the proportions in which elementary substances existed in a compound. As we saw, using these units demanded the assumption that intermediate non-integer values for the proportions did not exist, an assumption which, if true, would greatly facilitate the compositional identification of these substances, because the resulting *discontinuities* in the space of possible compounds eliminated a large number of possibilities.[71]

Evidence for discontinuous proportions came from a variety of sources, neither one sufficient to compel assent. If dividing weight percentages by the unit corresponding to the composing elements yielded an integer number that was considered confirming evidence, but in practice the actual numbers demanded rounding; this could be done by rounding up or down.[72] Another source of evidence was the behavior of the compound in various chemical reactions, in the course of which some elementary components remained together more often than others, but there were often a variety of ways in which this could be done depending on the chemical reaction used as a reference. So the rule of evidence that evolved was one of *convergence by mutual adjustment* of the different factors involved: if the numerical results of quantitative analysis could be reconciled with the decomposition products of chemical reactions, and if these could be reconciled with the way in which a substance tended to break down into pieces, the mutual fit of the different lines of evidence could itself be considered a source of evidence.[73] The justification for this rule was that even though none of the separate lines of evidence was compelling by itself, and even though phenomena, instrumentation, and cognitive tools were all treated as potentially modifiable, there were only a few ways in which the modifications could be made to yield mutually compatible results.[74]

A different kind of rule of evidence was needed to achieve the goal of *discovering the factors determining the properties of compounds*. The chemical nature of the ingredients and their relative proportions had already been reliably established as relevant factors by the previous two lines of research, but then the phenomenon of isomerism showed them to be insufficient: isomers have the same elements in the same proportions and yet display different properties. An additional causal factor could be the *spatial arrangement* of components, but adding this involved going beyond what was directly observable. This implied that generalizations about spatial structure could not be reached using induction, forcing chemists to propose a hypothetical cause and then test its consequences, that is, forcing them to use the hypothetico-deductive method. There had been tension between these two methodologies from the start of natural philosophy, because each was linked to a different research goal: the first aimed at creating compact descriptions of a phenomenon, the second at discovering the causes that produced the phenomenon. The first goal ("saving the phenomena") had been endorsed by the most prominent natural philosophers, becoming a widely accepted *explicit goal*, but the search for causes continued to exist alongside the first as an *implicit goal*.[75] This methodological disagreement complicated the process of arriving at new rules of evidence through argumentation, but this was nevertheless possible because the consequences derived from hypotheses could be treated as *testable predictions*.[76] Structural formulas could be used to predict the number of isomers that different substances should have and there was little dispute on how to produce evidence for this: synthesize the predicted isomers. On the other hand, correct predictions were not enough to compel the assent of anti-atomists, since they could accept the formulas as convenient isomer-counting devices, having no ontological consequences.[77] As in the previous two examples, this left plenty of room for local underdetermination.

Let's give one final example to illustrate the limits of evidence to create consensus. When physical chemists set out to study what happens when a compound like salt is dissolved in a substance like water, they introduced an unfamiliar new tool into chemistry: mathematical models. These were problematic in several respects. First of all, they necessarily involved idealizations and these eliminated what interested traditional chemists. Idealizing away all interactions between solvent and solute not only eliminated the *variation* in the properties of substances, but it also left out all chemical processes: when salt becomes dissociated into its components (sodium, chlorine) as it dissolves in water, the traditional explanation was that the affinity between water and the components of salt was greater than the affinity between sodium and chlorine for each other. A mathematical model of dissociation that did not take affinity into account was not a relevant model at all.[78] Second, the new tools inverted the relation between phenomena and evidence. In the case of isomerism, for example, the phenomenon came first and the hypothetical models to explain it, as well as their prediction of the number of possible isomers, came later. But in the present case, the ideal phenomenon and its model came first, and the laboratory phenomenon (a solution that approximated infinite dilution) came later. This meant that even though predictions made on the basis of the model were backed by the evidence, the significance of the predictions could be questioned. And there were other predictions, made on the basis of non-mathematical models, that also had supporting evidence.[79]

These four examples illustrate different degrees of difficulty in the formation of consensus. The recovery of a compound's properties after they had disappeared under analysis was compelling evidence for the qualitative composition of inorganic substances. The slow convergence of units and formulas, on the one hand, and data from quantitative analyses of substances, and from the products of the reactions into which these participated, on the other, supplied a less direct line of evidence than the analysis/synthesis cycle. But together with the assumption that formulas and data can be mutually adjusted in only a few ways, it was enough to justify statements about quantitative composition. The causal role of spatial arrangement was harder to justify. Successful predictions about the number of isomers that a given compound could have carried weight, but not enough to compel belief in invisible entities. Finally, attempts to figure out what happens to dissolved compounds, and how to distinguish compounds (with definite proportions) from mixtures (with indefinite proportions), led to the confrontation of very different approaches, each with its own evidence rules. In this case, a communication breakdown, and the reaching of consensus through non-cognitive factors, was a real possibility. It will be convenient to separate the first three examples, in which arguments can lead to partial agreement, from the fourth, in which temporarily irresolvable controversies are involved. The question of the role of authority in the formation of consensus takes a different form in these two situations.

We may distinguish three forms of authority in a scientific field: the cognitive authority of consensus practice; the personal authority of individual practitioners; and the professional authority of the field itself, that is, the kind of authority that allowed nineteenth-century chemists to play the role of expert witnesses, government inspectors, and industry consultants.[80] One way of incorporating personal authority is through the concept of a *reference group*. This denotes a group of practitioners, real or imagined, whose personal practice is used by other members of the community to judge their own performance and to guide their career choices.[81] If the practices of the reference group are representative of the state of the field at any one time then it can help spread innovations and improvements to the less prominent members. Thus, if chemists in the 1750s used Macquer as their reference, or chemists in the 1830s used Berzelius, they would be spreading best practices and helping forge a new consensus.[82] In this case, personal authority is earned through concrete *exemplary achievements*, and these can serve as a source of inspiration and a guide to practice for other members of the community. On the other hand, even when the authority of influential chemists is based on merit, it can also serve to preserve cognitive tools that may have outlived their usefulness, as illustrated by Macquer's attempt to save phlogiston by redefining it as the matter of light, or by Berzelius's continued defense of his model of the binary composition of organic substances. Moreover, even when a reference group focuses the attention of the community on legitimate topics, it may for that same reason steer them away from other worthy pursuits. This may explain why the rise of physical chemistry took place in the periphery where practitioners were not subject to the gravitational pull of the reference groups in Berlin or Paris.[83]

A similar point applies to the forums of discussion—conferences, journals, meetings—in which critical debate is carried out. These forums can help define what is being debated, through the publication of essays assessing the merits of different alternatives, allowing the expression of rival points of view, and moderating polemics. But as in the case of other institutions, the personal authority of the editor of a journal, the head of a university department, or the director of a professional organization can be abused: worthy papers are denied publication, worthy candidates denied an academic post, and worthy contenders denied a prize or an honor.[84] Can the practitioners controlling access to a forum interfere and distort the process of consensus formation? Two conditions must be met to answer this question correctly: the institutional organizations forming the social infrastructure of a given field (learned academies, professional societies, publishers of journals and textbooks) must be conceived as linked to each other by relations of exteriority; and the genealogical link between masters and disciples conceived as involving *redistributions of authority across generations*.[85] If these two conditions are met then the population of organizations will tend to include at any one time older ones co-existing with newer ones, giving younger scientists a choice: they can decide to pursue a career chasing after prestigious positions in existing organizations; to join a new organization with a fresher outlook; or even to set up a rival one, as when young practitioners decide to create their own journal rather than fighting for control of an established one.[86]

A similar conclusion can be reached about reference groups: they can be, and often are, challenged. Every generation of students is taught not only the consensus practice but also a critical attitude towards it, and those that have acquired the status of spokesmen for the field must be prepared to defend it against this criticism, and to do so in the name of the field, by invoking shared values and collectively accepted arguments, not their personal opinion.[87] Personal authority tends to display a high degree of *pluralism*, and the plurality of sources of authoritative opinion and criticism can prevent concentrations of power from distorting the formation of consensus.[88] This leaves only irresolvable controversies as a possible obstacle to the reasoned passage from personal to consensus practice. The case just mentioned, the clash of physical chemistry with more traditional approaches to the study of reactions in solution, can serve as an illustration. This controversy offered the perfect conditions for a complete breakdown in communication, and the consequent role for political alliances in deciding its outcome. In its early stages, the two sides did tend to argue at cross purposes, one side using evidence from experiments at high dilution while the other offered counterexamples from experiments at higher concentrations. And the two rival leaders did behave as if there was an unbridgeable gap, refusing to recognize any common ground: Ostwald often dismissed the claims of his adversaries, entirely ignoring them when writing his textbook, and on the other side the British chemist Henry Amstrong represented Ostwald's position as a monolithic entity in which concepts, statements, and explanations had to be accepted or rejected as a whole.[89]

But like other debates in the history of chemistry, the controversy had a temporal structure: arguments evolved in response to criticisms; new evidence was produced and assessed; and intermediate positions were developed. Journals and chemical societies played a constructive role, organizing encounters between the rival factions, reporting both sides of the discussion, and assessing the cognitive costs and benefits of each position. The British Association for the Advancement of Science organized a grand meeting in 1890 at Leeds, attended by both factions. Neither side prevailed but their positions did come closer together. The Faraday Society and the American Electrochemical Society followed the debate. Journals like *Philosophical Magazine* and *Nature* published information on both sides.[90] In addition, some participants exerted a moderating force, bringing the exchange back from polemics to serious debate, promoting goodwill, and smoothing out misunderstandings.[91] In time, the two rival leaders tired of the debate and withdrew, leaving the participants occupying the middle ground to forge a compromise: liquid solutions can contain both dissociated ions *and* hydrate compounds, a position closer to the truth than any of the original ones.[92]

This mechanism of conflict resolution, the formation of plausible intermediate positions between extreme rivals, is clearly not universal. Other mechanisms exist, such as the common search of the space of alternatives, and the progressive pruning of less viable candidates, that we discussed in Chapter 1, a collective instantiation of eliminative induction.[93] More detailed historical analyses of actual controversies may reveal yet other ways in which the resolution of conflict by the imposition of authority can be prevented. But even in those cases in which an authoritarian outcome has not been avoided, we must assess the extent to which those who imposed it managed to make it endure past their own generation, or prevail beyond their own country. If these remarks are roughly correct, and if the previous arguments about the benign role of conventions and the superficial role of myth, propaganda, and rhetoric is also close to the mark, we can accept that a scientific community is socially situated without compromising the integrity of the content it produces. To this tentative conclusion we may add those from previous chapters: that the cognitive tools shaping the personal practices of the members of a scientific community can improve, as measured by each tool's own standards, and that cognitive innovation is a contingent but real feature of the history of scientific fields. The overall argument supports the view that scientific communities are reliable producers of cognitive content, even if this production is always risky and its product fallible. Moreover, none of this retracts from the remarks made at the start of this book: that the population of scientific fields is divergent, that is, it is not converging on a final theory; and that the domain of most fields becomes more complex with time, presenting practitioners with a moving target and precluding any dream of a final truth.

REFERENCES

Introduction

1 Stephen Toulmin. *Human Understanding. The Collective Use and Evolution of Concepts* (Princeton, NJ: Princeton University Press, 1972), p. 141.

Toulmin's book can be considered the first contemporary attempt at creating a *realist* social epistemology. In addition, Toulmin was the first philosopher to use a populational approach to scientific fields, stressing the historicity of disciplinary boundaries, much like those demarcating different biological species. In Toulmin's model, cognitive content exists as a population of concepts. We will extend this to a variety of cognitive tools: not only concepts, but true statements, significant problems, and explanatory and taxonomic schemas. This is less of a departure than it seems because the other cognitive products, such as problems, are also involved in Toulmin's account. For example, in the interplay between conceptual variation and critical intellectual selection, innovations are selected on the basis of the role concepts play in the posing and solving of problems (ibid., p. 207).

2 Dudley Shapere. "Scientific Theories and Their Domains." In *The Structure of Scientific Theories*. Edited by Frederick Suppe (Chicago: University of Illinois Press, 1977), p. 518.

Shapere introduced the concept of a *domain* to give a less vague definition of the subject matter of a given field, a subject matter that may have a lesser or greater degree of order depending on the historical period under study. Shapere thought of this concept as replacing the old distinction between observation and theory, since a domain contains both observable phenomena as well as the theoretical classifications that give order to these phenomena. However, as sympathetic commentators have noted, Shapere's own treatment is unclear as to whether domains are collections of phenomena, linguistic descriptions of phenomena, or sets of problems raised by phenomena.

Thomas Nickels. "Heuristics and Justification in Scientific Research: Comments on Shapere." In ibid., p. 584.

3 Ian Hacking. *Representing and Intervening* (Cambridge: Cambridge University Press, 1983), pp. 224–8.

Hacking's treatment of the concept of phenomenon is compatible with Shapere's goal of replacing the old dichotomy between observation and theory. Take, for example, his requirement that any phenomenon be remarkable or noteworthy. In the astronomical case, the wandering stars, that is, the loopy trajectories of planets like Mars, meet this requirement by their dramatic contrast with the relatively fixed positions of the stars. But in the laboratory, an observed effect is often considered noteworthy or instructive relative to an existing theory (ibid., p. 226). On the display of phenomena as public entities see also: Stephen Toulmin. *Human Understanding*. Op. cit., pp. 195–8.

4 Joachim Schummer. "Towards a Philosophy of Chemistry." *Journal for General Philosophy of Science*, Volume 28, Number 2 (Netherlands: Springer, 1997), p. 327.

5 Philip Kitcher. *The Advancement of Science. Science without Legend, Objectivity without Illusions* (New York: Oxford University Press, 1993), pp. 74–86.

This list—concepts, statements, problems, explanatory schemas, and taxonomic schemas—is adapted from the list given by Kitcher. His list includes some items (exemplars) left out from the current list, while it excludes other items (taxonomies) that are added here to articulate his ideas with those of Shapere. The list must be considered open, so that further components may be added: chemical formulas, mathematical models, computer simulations.

6 Frederick Suppe. "Afterword – 1977." In *The Structure of Scientific Theories*. Op. cit., p. 686.

Of the several possible interpretations of Shapere's notion noted in note 2, above, Suppe chooses that of a domain as an ordered *body of information*. In this book a more embodied version is favored: a domain as an ordered collection of phenomena. But it is clear that when we articulate a domain with a community of practitioners, what flows from one to the other is indeed information: there will typically be a signal or information channel (natural or technological) connecting the two, carrying signs (bits of information) from one to the other.

7 Yakov M. Rabkin. "Uses and Images of Instruments in Chemistry." In

Chemical Sciences in the Modern World. Edited by Seymour H. Mauskopf (Philadelphia, PA: University of Pennsylvania Press, 1993), pp. 28–31.

8 Stephen Cole. *Making Science. Between Nature and Society* (Cambridge, MA: Harvard University Press, 1995), p. 230.

9 Philip Kitcher. *The Advancement of Science.* Op. cit., p. 88.

The distinction between personal and consensus practices is Kitcher's, but his terminology has been changed. Kitcher uses the term "individual practice" not "personal practice." In this book the term "individual" is used to define an ontological status, the status of a unique and singular historical being, and it is therefore used as a qualifier for all kinds of entities: individual persons, individual communities, individual organizations, individual scientific fields.

10 Stephen Cole. *Making Science.* Op. cit., pp. 82–96.

Evidence for the existence of variation in personal practices comes from a statistical analysis of peer-reviewed grant applications. Another source of evidence could have been a survey of laboratories, conducted by providing the staff of each research organization with an appropriate questionnaire. But analyzing the results of grant evaluations seems more appropriate because it engages working scientists as they carry one of their professional responsibilities, and because it avoids artifacts in the design of the questionnaire. The results of one detailed quantitative study show that the degree of consensus in a community over the content of cutting-edge research is extremely low.

11 Thomas S. Kuhn. *The Structure of Scientific Revolutions* (Chicago: University of Chicago Press, 1970), pp. 136–43.

Kuhn argues that textbooks are to be blamed for the invisibility of scientific revolutions by their propensity to depict the history of science as continuous and cumulative. We do not have to accept the distinction between normal and revolutionary science, however, to agree that the presentation of history in textbooks is very unreliable. If instead of the distinction between normal and revolutionary science we use the one between consensus and frontier science, then a more nuanced critique of these teaching aids can be made. Textbooks include what has become consensus in a more or less uniform way, but there is variation among them on the rate at which frontier science is included. See: John Hedley Brooke. "Introduction: The Study of Chemical Textbooks." In *Communicating Chemistry: Textbooks and Their Audiences, 1789–1939.* Edited by Anders Lundgren and Bernadette Bensaude-Vincent (Canton: Watson Publishing, 2000), pp. 5–6.

Chapter 1: Classical Chemistry

A multiplicity of cognitive tools

1 Jaap van Brakel. *Philosophy of Chemistry* (Leuven: Leuven University Press, 2000), pp. 71–3.

2 Ibid., pp. 124–5.

Many philosophers of science tend to believe that thermodynamics was reduced to statistical mechanics and that, therefore, temperature was reduced to average kinetic energy. But as van Brakel argues, this is true only for very low-density gases (those that behave like ideal gases), not for solids, plasmas, the vacuum, or even gases in molecular beams. A more detailed account of the difficulties of reducing thermodynamics to statistical mechanics can be found in: Lawrence Sklar. *Physics and Chance. Philosophical Issues in the Foundations of Statistical Mechanics* (Cambridge: Cambridge University Press, 1995), pp. 345–54.

3 Joachim Schummer. "Towards a Philosophy of Chemistry." *Journal for General Philosophy of Science*, Volume 28, Number 2 (Netherlands: Springer, 1997) p. 327.

4 Ibid. pp. 316–317.

The hierarchy as just presented is a modern version. In the early 1700s a chemist like Gustav Stahl would have formulated it like this: at the bottom there were principles; then mixts composed of principles; then compounds made of mixts; and then finally aggregates made of compounds. See: Mi Gyung Kim. *Affinity, That Elusive Dream. A Genealogy of the Chemical Revolution* (Cambridge, MA: MIT Press, 2003), pp. 171, 195.

5 William Whewell. *History of Scientific Ideas*. Vol. 2 (London: John W. Parker and Son, 1858), p. 22.

Whewell quotes Stahl who gives a description of this phenomenon as if it was already routine by the time he was writing in 1697.

6 Bernadette Bensaude-Vincent and Isabelle Stengers. *A History of Chemistry* (Cambridge, MA: Harvard University Press, 1996), pp. 69–70.

7 R. G. W. Anderson. "The Archaeology of Chemistry." In *Instruments and Experimentation in the History of Chemistry*. Edited by Frederic L. Holmes and Trevor H. Levere (Cambridge, MA: MIT Press, 2000), pp. 12–13.

8 Trevor H. Levere. *Affinity and Matter. Elements of Chemical Philosophy, 1800–1865* (Oxford: Clarendon, 1971), p. 35.

9 Stephen Toulmin and June Goodfield. *The Architecture of Matter* (Chicago: University of Chicago Press, 1982), p. 202.

10 Maurice Crosland. "Slippery Substances: Some Practical and Conceptual Problems in the Understanding of Gases in the Pre-Lavoisier Era." In *Instruments and Experimentation in the History of Chemistry*. Op. cit., p. 84.

11 Trevor H. Levere. "Measuring Gases and Measuring Goodness." In *Instruments and Experimentation in the History of Chemistry*. Op. cit., p. 108.

Levere claims that between the first identification of gases as distinct substances by Joseph Black in the 1750s and the end of the century, balances increased in precision 100 times.

12 John Hudson. *The History of Chemistry* (New York: Chapman & Hill, 1992), pp. 70–1.

The expression is Antoine Lavoisier's, who is generally credited with the codified definition of the concept of "element" as the limit of chemical analysis.

13 Robert Siegfried. *From Elements to Atoms. A History of Chemical Composition* (Philadelphia, PA: American Philosophical Society, 2002), p. 89.

As early as 1699, Wilhelm Homberg was using the analysis–synthesis cycle to establish qualitatively the composition (and hence the identity) of bodies, a practice that was crucial in the development of the concept of "pure substance."

14 Alistair Duncan. *Laws and Order in Eighteenth-Century Chemistry* (Oxford: Clarendon, 1996), pp. 31–7.

15 Ibid., pp. 185–7.

The empirically revealed complexity of affinity relations was "built into" the concept by Pierre-Joseph Maquer and Antoine Baume through classifications of different types of affinity.

16 Mi Gyung Kim. *Affinity*. Op. cit., pp. 134–8.

The original affinity table was published in 1718 by Etienne-Francois Geoffroy.

17 Ibid., p. 141.

18 Alan Garfinkel. *Forms of Explanation* (New Haven, CT: Yale University Press, 1981), pp. 38–9.

A more formal treatment of problems modeled as "why" questions is given by Wesley Salmon. In Salmon's model the subject and predicate together are called the "topic" and the alternatives the "contrast class." These two components must be supplemented by a relevance relation, specifying what counts as a significant answer to the question, whether the answer must specify a cause, for example, or whether it can limit itself to specifying a function. Finally, presuppositions are listed as a separate component. See: Wesley C. Salmon. *Scientific Explanation and the Causal Structure of the World* (Princeton, NJ: Princeton University Press, 1984), pp. 102–110.

19 William H. Brock. *The Chemical Tree. A History of Chemistry* (New York: W.W. Norton, 2000), p. 82.

20 Mi Gyung Kim. "The 'Instrumental' Reality of Phlogiston." *Hyle International Journal for the Philosophy of Chemistry*. Volume 14, Number 1 (Berlin: Hyle Publications, 2008), p. 31.

The term "phlogiston" was the name given by Stahl to the Sulphur principle.

21 Alistair Duncan. *Laws and Order*. Op. cit., p. 190.

This eighteenth-century terminology can be easily changed to the one used today. The question would then be phrased like this: Why does a reaction of potassium aluminum sulphate and potassium hydroxide yield potassium sulphate instead of another compound?

22 Philip Kitcher. *The Advancement of Science. Science without Legend, Objectivity without Illusions* (New York: Oxford University Press, 1993), p. 82.

The concept of an *explanatory schema* is Kitcher's. He defines it as a set of *schematic sentences* (sentences in which some of the nouns are replaced by dummy variables) and a set of filling instructions to replace the variables with words. He acknowledges that the schemas may not appear as such in the practice of any scientist, but argues that they do capture implicit reasoning patterns actually used by scientists.

23 Robert Siegfried. *From Elements to Atoms*. Op. cit., p. 72.

24 An example of a problem of this type was: Why does potassium sulphate and calcium nitrate react to produce potassium nitrate and selenite instead of other compounds? The explanatory device was a diagram that placed the two reacting compounds on the left and right sides next to vertical curly brackets embracing their ingredients: potassium hydroxide and sulphuric acid for the first, nitric acid and calcium oxide for the second. The components were placed in the diagram so that those with greater affinity were next to one another in the upper or the lower part of the diagram. Finally, horizontal curly brackets embraced the two pairs of liberated components, with the resulting new compounds placed on the other side of the bracket. See: Alistair Duncan. *Laws and Order*. Op. cit., pp. 199–211.

Duncan does not see these diagrams as explanatory. They serve to display the relationships between the components of the reactants, and show that these components remain unchanged as they enter into new compounds, but they do not give the *cause* of the transformation. It can be argued, however, that affinities, as emergent capacities to combine, are partial *macro*-causes of the formation of new compounds, although affinities themselves need further explanation in terms of *micro*-causes, such as the breaking of bonds and the formation of new ones. Perhaps what Duncan has in mind is that to the extent that other important macro-causes, such as temperature and saturation, are left out of the diagrams, they cannot be considered to be *complete* explanations.

25 Hillary Putnam. "The Meaning of 'Meaning'." In *Mind, Language, and Reality* (London: Cambridge University Press, 1979), pp. 227–8.

Actually, this test would leave undetermined whether the sample was gold or platinum. Further tests would be needed to break the tie. The philosopher who first included expert intervention as part of what fixes reference was Hillary Putnam, in his thesis of the "division of linguistic labor." A more recent kindred approach can be found in: Jaap van Brakel. *Philosophy of Chemistry*. Op. cit., pp. 103–5.

26 Robert Siegfried. *From Elements to Atoms*. Op. cit., p. 70.

27 Bernadette Bensaude-Vincent and Isabelle Stengers. *A History of Chemistry*. Op. cit., pp. 55–6.

The authors give the example of Duhamel's analysis of sea salt to isolate the alkali (sodium carbonate) that is one of its components. When combined with "marine acid" (hydrogen chloride in solution) it recreated sea salt. Thus, a substance traditionally produced by extraction (sea salt) was shown to be producible in the laboratory. In our terms, the referent of "sea salt" was not fixed any longer by reference to extraction from sea water but by the analysis–synthesis cycle, and this constituted a clear improvement.

28 Mi Gyung Kim. *Affinity*. Op. cit., p. 191.

Kim describes the work of Guillaume-Francois Rouelle, who in the middle of the century set out to classify all the known neutral salts using crystal shape and solubility as identifying properties and dispositions. Later in the century, Carl Scheele was routinely using crystal shape, boiling point, and solubility. See: William H. Brock. *The Chemical Tree*. Op. cit., p. 174.

29 Robert E. Schofield. *The Enlightened Joseph Priestley. A Study of His Life and Work From 1773 to 1804* (University Park, PA: Pennsylvania State University Press, 2004), p. 186.

30 Joseph Priestley is credited with the discovery of oxygen, despite his misdescription of it as "air without phlogiston." See discussion of the different ways of fixing reference in this historical case in: Philip Kitcher. *The Advancement of Science*. Op. cit., pp. 99–102.

31 Richard L. Kirkham. *Theories of Truth* (Cambridge, MA: MIT Press, 1992), pp. 20–31.

Kirkham distinguishes three different goals or projects that theorists of truth can have: the metaphysical project (what is truth?), the justification project (what other property correlates with truth?) and the speech-act project (what is achieved or performed when making claims to truth?).

32 Ibid., pp. 216–20.

These are both "correspondence" theories of truth. But the relation of correspondence may be conceived as an isomorphism, as Bertrand Russell did, or as a correlation through conventions, as J. L. Austin did.

33 Ibid., pp. 120–9.

34 Robert Siegfried. *From Elements to Atoms*. Op. cit., pp. 97–8.

In modern terminology this would be phrased like this: The statement "Sodium sulfate is composed of sulfuric acid and sodium hydroxide" is true if sodium sulfate is composed of sulfuric acid and sodium hydroxide.

Accepting a statement like this amounts to accepting Alfred Tarsky's conception of truth, a correspondence theory in which truth is defined as *satisfaction*. But strictly speaking, Tarsky's conception applies only to formal languages like the predicate calculus, not to natural languages. And, in Tarsky's own version, satisfaction itself could not be explained in terms of simpler relations like reference. See: Richard L. Kirkham. *Theories of Truth*. Op. cit., pp. 161–2.

In what follows it will be assumed that a version of Tarsky's conception exists that can be applied to natural languages and that satisfaction can be reduced to reference.

35 All we need to reject is that the relations between statements are such that they define their very identity. Relations that constitute the identity of what they relate are referred to as *relations of interiority*, while those that do not are *relations of exteriority*. In this book all relations (with the possible exception of the constitutive conventions discussed in the final chapter) are of the latter type: the relations between components of a domain; those between different cognitive tools; and those between statements.

36 Mi Gyung Kim. *Affinity*. Op. cit., p. 192.

This generalization is credited to Rouelle.

37 Frederick L. Holmes. *Antoine Lavoisier: The Next Crucial Year* (Princeton, NJ: Princeton University Press, 1998.), p. 7.

Historians usually credit Lavoisier with codifying this conservation principle, but they acknowledge that it existed before him in tacit form.

38 Frederick Lawrence Holmes. *Lavoisier and the Chemistry of Life* (Madison, WI: University of Wisconsin Press, 1985.), p. 229.

39 Ibid., pp. 218, 234.

This meant that results had to be adjusted to assumptions and assumptions to results. This did not involve vicious circularity, however, because the numbers that were sought were those on which several experiments converged, but it did make the numbers far from certain.

40 Mi Gyung Kim. *Affinity*. Op. cit., p. 259.

The extension to 50 columns in 1775 was done by Torbern Bergman. The 59-column version would have involved 30,000 experiments, according to Bergman (ibid., p. 268).

41 Alistair Duncan. *Laws and Order*. Op. cit., pp. 137–8.

42 Bernadette Bensaude-Vincent and Isabelle Stengers. *A History of Chemistry*. Op. cit., pp. 71–2.

The objective phenomenon itself, the spontaneous production of a salt lake with banks covered with soda, was observed by Claude-Louis Berthollet in Egypt when accompanying the expedition led by Napoleon in 1800.

43 Alan Garfinkel. *Forms of Explanation*. Op. cit., pp. 30–2.

44 G. C. Donington. *A Class-book of Chemistry* (London: Macmillan, 1920), pp. 260–1.

In 1799, Louis José Proust investigated the compounds that copper, iron, and other metals formed with oxygen and sulphur (oxides, sulfides), and found they combined in definite proportions.

45 Alistair Duncan. *Laws and Order*. Op. cit., p. 164.

46 Ibid., pp. 194–6.

A statement like PW-2' appears in the "laws" of affinity codified by
Antoine Francois Fourcroy and Guyton de Morveau.

47 Philip Kitcher. *The Advancement of Science*. Op. cit., p. 92.

48 Thomas S. Kuhn. *The Structure of Scientific Revolutions*. Op. cit.,
pp. 150–2.

49 Philip Kitcher. *The Advancement of Science*. Op. cit., pp. 256–7.

The three features of this model—the escape tree, the collective search
of the space of alternatives, and the constructive role of critics—are
proposed by Kitcher as a means to re-conceptualize controversies.

50 Ibid., pp. 200–1.

51 Ibid., pp. 344–5.

Kitcher argues that when no participant in a controversy knows what the
best outcome should be, it is better to spread the risks around rather than
betting it all on a single winner

52 Larry Laudan. *Progress and Its Problems* (Berkeley, CA: University of
California Press, 1977), pp. 148–50.

53 Thomas S. Kuhn. *The Structure of Scientific Revolutions*. Op. cit., p. 118.

54 Mi Gyung Kim. *Affinity*. Op. cit., pp. 386–8.

55 Ibid., pp. 338–9.

56 Maurice Crosland. "Slippery Substances." Op. cit., pp. 98–100.

57 Larry Laudan. *Progress and Its Problems*. Op. cit.

Laudan would characterize the replacement of "air" and "airs" by "gas" as
an example of solving a *conceptual problem*. He distinguishes empirical
from conceptual problems. Empirical problems are substantive questions
about the content of the domain (ibid., p. 15), although stating their
presuppositions and relevant contrasts involve the use of concepts.
Conceptual problems, on the other hand, arise either because of logical
contradictions or because vaguely defined notions lead to ambiguities
or circular definitions (ibid., p. 49). Conceptual problems are created by
relations of incompatibility, entailment, reinforcement, or implausibility,
between concepts belonging to different disciplines, even non-scientific
ones like philosophy (ibid., p. 54). Here we refer to these problems as
"difficulties" or "obstacles" to distinguish them from empirical problems
modeled as "Why" questions.

58 Philip Kitcher. *The Advancement of Science*. Op. cit., p. 255.

This situation is traditionally defined as the *underdetermination* of theoretical choices by the available evidence. Kitcher agrees that it can occur but that it is *only temporary*, lasting for part of a debate but being eliminated as new evidence is produced, or as the participants become aware of the costs of abandoning the truth of older statements.

59 The seven major participants were: Guyton de Morveau, Antoine Francois Fourcroy, Claude-Louis Berthollet, Antoine Lavoisier, Richard Kirwan, Henry Cavendish, and Joseph Priestley.

60 Mi Gyung Kim. *Affinity*. Op. cit., p. 315.

The landmarks of pneumatic chemistry are roughly these: fixed air was isolated by Joseph Black in 1756, who showed that unlike ordinary air it would not support combustion. Inflammable air was isolated by Cavendish in 1766. Priestley isolated 20 new airs between 1770 and 1800, including dephlogisticated air in 1775. Lavoisier was ignorant of these results in 1772—*the air that was fixed during calcination in his hypothesis was ordinary air*—but he had become the recognized French expert in pneumatic chemistry by early 1774.

61 Philip Kitcher. *The Advancement of Science*. Op. cit., pp. 247–51.

The situation described in the main text is traditionally defined as involving *underdetermination*. Kitcher agrees that underdetermination may occur in two circumstances. First, there may be underdetermination that is *only temporary*, lasting for part of a debate but being eliminated as new evidence is produced, or as the participants become aware of the costs of abandoning the truth of older statements (ibid., p. 255). Or there may be another kind of underdetermination: *several rival ways of assessing the costs* may exist, leading to different conclusions as to how to fit a new statement into the old corpus (ibid., p. 252) Either way, there is no general philosophical argument to handle these cases: one must look at the historical evidence, one individual controversy at a time.

62 Mi Gyung Kim. *Affinity*. Op. cit., pp. 241–2.

It was Guyton de Morveau who established the general existence of the phenomenon, and who first suggested links to reduction and acidification, thereby providing a framework for the debate.

63 No participant ever questioned the role of affinity, so they all shared statement 1. Difficulties in accounting for the affinities of oxygen, for example, could be used as counterarguments accepted by both sides. Thomas Kuhn is mistaken when he asserts that affinity theory was part of a paradigm that was left behind at the start of the nineteenth century. In fact, the concept of affinity continued to be used and explored for the next hundred years: Thomas S. Kuhn. *The Structure of Scientific Revolutions*. Op. cit., p. 132.

Statement 2 (the conservation of weight) had been accepted implicitly by chemists throughout the century as the main assumption justifying the use of the balance in the laboratory by chemists like Homberg and Geoffroy. See: Robert Siegfried. *From Elements to Atoms*. Op. cit., pp. 87, 93.

64 Frederick L. Holmes. *Antoine Lavoisier*. Op. Cit., pp. 13–28.

The challenges were by made Antoine Lavoisier, starting in 1772. But Lavoisier knew that influential members of the French chemical community (principally Macquer) would resist a frontal attack on the concept of phlogiston, so he was careful not to attempt that until he had gathered enough evidence and allies.

65 William H. Brock. *The Chemical Tree*. Op. Cit., p. 103.

Lavoisier observed the emission of air in the reduction of litharge (one form of lead oxide), a phenomenon that seemed so remarkable to him that towards the end of 1772 he deposited a sealed account of it in the archives of the Academy to be opened in May 1773, once he had had the time to confirm it.

66 Mi Gyung Kim. *Affinity*. Op. cit., pp. 314–15.

67 Ibid., pp. 244–5. This proposal was made by Guyton de Morveau.

68 In a retrospective account written in 1792, Lavoisier himself offered ways of blocking this added branch. And Priestley himself, the very last hold-out, had rejected this way out as early as 1775. Both are quoted in: Philip Kitcher. *The Advancement of Science*. Op. cit., pp. 277–9.

69 Ibid., p. 283.

70 William H. Brock. *The Chemical Tree*. Op. cit., pp. 109–10.

The synthesis of water was first observed by Priestley, and later communicated to Cavendish, in 1781. The latter repeated the experiment and reported it to the Royal Society in 1784. The attempt to save the elementary status of water was also Cavendish's.

71 Andrew Pile. "The Rationality of the Chemical Revolution." In *After Popper, Kuhn, and Feyerabend*. Edited by Robert Nola and Howard Sankey (Dordrecht: Kluwer Academic Publishers, 2000), p. 111.

Pile gives the example of ordinary sulphur. In one explanation it was a compound formed by reactions that progressively added phlogiston, starting with sulphuric acid, passing through sulphureous acid, and ending as sulphur. In the rival explanation, sulphur was elementary, and became sulphureous acid, and then sulphuric acid, as a result of reactions in which the amount of oxygen increased.

72 Mi Gyung Kim. *Affinity*. Op. cit., p. 147.

Kim argues that the status of phlogiston itself was analytically contingent. The term "phlogiston" had a hypothetical referent with properties and dispositions that seemed well established: it headed one of the columns of the original affinity table as the only substance that could displace the strongest acid (sulfuric acid) from its union with the strongest alkali (potassium carbonate). On the other hand, it had resisted all efforts at isolating it and purifying it, and this was bound to count against it as techniques of analysis improved.

73 Ibid., pp. 335–9.

74 Thomas S. Kuhn. *The Structure of Scientific Revolutions*. Op. cit., p. 148.

Kuhn asserts that when Lavoisier won the controversy this led not only to the loss of a permissible problem but also of its solution. Both of these statements are false.

75 Mi Gyung Kim. *Affinity*. Op. cit., p. 411.

Berthollet was the first to reveal holes in the oxygen model of acids, a fact that demonstrates that among the defenders of oxygen there could be partial disagreements. In 1787 he performed the first analysis of an acid (prussic acid) that had no oxygen in it, but this result was ignored.

76 Robert Siegfried. *From Elements to Atoms*. Op. cit., p. 192.

77 William H. Brock. *The Chemical Tree*. Op. cit., p. 108.

From personal to consensus practice 1700–1800

1 Ursula Klein and Wolfgang Lefévre. *Materials in Eighteenth-Century Science. A Historical Ontology* (Cambridge, MA: MIT Press, 2007), p. 16.

2 Ibid., p. 34.

3 Mi Gyung Kim. *Affinity, That Elusive Dream. A Genealogy of the Chemical Revolution* (Cambridge, MA: MIT Press, 2003), p. 85.

The last statement, using the term "middle salt" or "salty salt," was treated as an axiom or indubitable truth by 1701.

4 Ibid., p. 55.

These problems were recognized as among the most difficult questions to be resolved by Nicolas Lemery in a textbook published in 1675.

5 Ursula Klein and Wolfgang Lefévre. *Materials in Eighteenth-Century Science*. Op. cit., pp. 41–4.

6 Mi Gyung Kim. *Affinity*. Op. cit., pp. 28, 84.

Etienne de Clave and Robert Boyle had proposed similar ideas in the previous century. The textbook in which chemistry was defined as the art of analyzing and synthesizing was published in 1702 by Wilhem Homberg.

7 Ibid., p. 35.

8 Ibid., p. 30.

9 Ibid., pp. 52–5.

The change can be documented by comparing textbooks. In the Nicaise Lefebvre textbook of 1660, distillation alone is the exemplar of correct analysis. In Lemery's 1675 textbook, solution analysis is already featured as an alternative.

10 Ursula Klein and Wolfgang Lefévre. *Materials in Eighteenth-Century Science*. Op. cit., pp. 137–8.

11 Ibid., pp. 141–6.

12 Mi Gyung Kim. *Affinity*. Op. cit., p. 61.

13 Ibid., pp. 171–2.

14 Bernadette Bensaude-Vincent and Isabelle Stengers. *A History of Chemistry* (Cambridge, MA: Harvard University Press, 1996), pp. 58–9.

15 William H. Brock. *The Chemical Tree. A History of Chemistry* (New York: W. W. Norton, 2000), pp. 81–2.

16 Robert Siegfried. *From Elements to Atoms. A History of Chemical Composition* (Philadelphia, PA: American Philosophical Society, 2002), p. 87.

17 Ibid., p. 88.

18 Mi Gyung Kim. *Affinity*. Op. cit., pp. 81–3.

19 Ibid., pp. 90–6.

20 Ibid., p. 103.

21 Ibid.

22 Alistair Duncan. *Laws and Order in Eighteenth-Century Chemistry* (Oxford: Clarendon, 1996), pp. 168–9.

23 Jan Golinski. "Fit Instruments: Thermometers in Eighteenth-Century

Chemistry." In *Instruments and Experimentation in the History of Chemistry*. Edited by Frederic L. Holmes and Trevor H. Levere (Cambridge, MA: MIT Press, 2000), pp. 190–3.

24 Alistair Duncan. *Laws and Order*. Op. cit., pp. 57–8.

25 Maurice Crosland. "Slippery Substances: Some Practical and Conceptual Problems in the Understanding of Gases in the Pre-Lavoisier Era." In *Instruments and Experimentation in the History of Chemistry*. Op. cit., pp. 86–90.

26 William H. Brock. *The Chemical Tree*. Op. cit., p. 78.

27 Mi Gyung Kim. *Affinity*. Op. cit., pp. 191–2.

28 Ibid., pp. 196–7.

29 Bernadette Bensaude-Vincent and Isabelle Stengers. *A History of Chemistry*. Op. cit., p. 62.

30 Mi Gyung Kim. *Affinity*. Op. cit., p. 204.

31 Ibid., pp. 206–7.

32 Alistair Duncan. *Laws and Order*. Op. cit., pp. 182–6.

The names were coined by Macquer's collaborator Antoine Baumé.

33 The textbook in question is *Elémens de Chimie*, published in two volumes, one dedicated to theory (1749), the other to practice (1751). His *Dictionnaire de Chimie* (1766), aimed at a more general audience that included apothecaries, dyers, and metallurgists, was translated into English, German, Italian, and Dutch, and provided a model for other chemical dictionaries and encyclopedias. See: William H. Brock. *The Chemical Tree*. Op. cit., pp. 272–3.

34 Pierre Joseph Macquer. *Elements of the Theory and Practice of Chemistry*. Vol. I (Edinburgh: Donaldson and Elliot, 1777), p. 2.

Macquer uses the term "secondary principles" to refer to salty and oily substances that are the products of analysis, but which are thought to be composed of even more basic principles.

35 Ibid., pp. 19–20.

36 Ibid., pp. 14–15.

37 Ibid., pp. 21–2.

38 Ursula Klein and Wolfgang Lefévre. *Materials in Eighteenth-Century Science*. Op. cit., p. 138.

39 Pierre Joseph Macquer. *Elements of the Theory and Practice of Chemistry*. Op. cit., p. 62 (Copper), p. 84 (Antimony), p. 94 (Zinc).

40 Ibid., p. 25 (Selenite), p. 29 (salts from Nitrous Acid), and p. 37 (Borax).

41 Ibid., pp. 12–13.

Macquer mentions the different names for this "universal affection of matter" and asserts that we know it as a real effect "whatever be its cause." He goes on to list some characteristics of affinity and to discuss the more complex cases known to him.

42 These statements are paraphrases of assertions that occur in the following pages: ibid., pp. 27, 22, 19, 33.

43 Alistair Duncan. *Laws and Order*. Op. cit., pp. 112–15.

44 Pierre Joseph Macquer. Op. cit., p. 160.

45 Ibid., p. 45.

46 Ibid., p. 47.

A possible solution, according to Macquer, was that calces possessed a third component that was absent from other vitrifiable Earths: the Mercury principle.

47 Mi Gyung Kim. *Affinity*. Op. cit., p. 209.

48 Andrew Pile. "The Rationality of the Chemical Revolution." In *After Popper, Kuhn, and Feyerabend*. Edited by Robert Nola and Howard Sankey (Dordrecht: Kluwer Academic Publishers, 2000), p. 111.

49 Robert Siegfried. *From Elements to Atoms*. Op. cit., pp. 112–13.

50 William H. Brock. *The Chemical Tree*. Op. cit., p. 77.

51 Jan Golinski. "Fit Instruments: Thermometers in Eighteenth-Century Chemistry." In *Instruments and Experimentation in the History of Chemistry*. Edited by Frederic L. Holmes and Trevor H. Levere (Cambridge, MA: MIT Press, 2000), p. 198.

52 The credit for the standardization goes to George Martine. Black admired his work: his students were made to copy Martine's table, comparing 15 different thermometer scales. See: Jan Golinski. Op. cit., p. 195.

53 Christa Jungnickel and Russsell McCormmach. *Cavendish* (Philadelphia, PA: American Philosophical Society, 1996), p. 157.

54 Robert Siegfried. *From Elements to Atoms*. Op. cit., pp. 153–5.

55 Mi Gyung Kim. *Affinity*. Op. cit., p. 259.

56 Ibid., pp. 267–8.

57 Bernadette Bensaude-Vincent and Isabelle Stengers. *A History of Chemistry*. Op. cit., pp. 69–70.

58 Alistair Duncan. *Laws and Order*. Op. cit., p. 138.

59 Ibid., p. 220.

60 Ibid., pp. 198–204.

Among the precursors of this method, Duncan discusses Black and Bergman. Later on, Antoine Fourcroy used a similar approach.

61 Ibid., p. 205.

62 Maurice P. Crossland. *Historical Studies in the Language of Chemistry* (New York: Dover, 1978), p. 164.

Bergman was, in fact, the first to advocate nomenclature reform and the first to suggest using the genus/species distinction as a template to build new names for substances. But although this suggests that Guyton was simply a follower of Bergman, Crossland convincingly argues that their interaction was a true collaboration.

63 Ibid., pp. 68–86.

64 Ibid., p. 161.

65 Mi Gyung Kim. *Affinity*. Op. cit., p. 241.

66 Christa Jungnickel and Russsell McCormmach. *Cavendish*. Op. cit., pp. 259–60.

67 William H. Brock. *The Chemical Tree*. Op. cit., pp. 104–6.

68 Christa Jungnickel and Russsell McCormmach. *Cavendish*. Op. cit., pp. 264–5.

Cavendish's "solution," on the other hand, carried cognitive costs, since his proposal conflicted with the idea that phlogiston had been finally isolated in the form of inflammable air.

69 Frederic L. Holmes. *Antoine Lavoisier: The Next Crucial Year* (Princeton, NJ: Princeton University Press, 1998), p. 41.

70 Ibid., p. 42.

71 Ibid., pp. 94–9.

72 Mi Gyung Kim. *Affinity*. Op. cit., pp. 326–9.

73 Ibid., p. 275.

74 Alistair Duncan. *Laws and Order*. Op. cit., p. 210.

Had Kirwan gone further and picked a single acid as a standard, using parts per weight as a unit of measure, and then tabulated the *equivalent weights* of saturating bases for that acid, he would have pioneered the practice that came to be known as "stoichiometry."

75 Ibid., p. 220.

76 Mi Gyung Kim. *Affinity*. Op. cit., p. 351

77 Ibid., pp. 351–4.

If this assertion is correct then the notion of an *emergent property* can be traced to Fourcroy, at least in its explicit or codified form. In implicit form, however, the concept was already implied by the compositional hierarchy of Stahl, and by arguments about the irreducibility of chemistry to natural philosophy. On Fourcroy's contribution see also: Alistair Duncan. *Laws and Order*. Op. cit., p. 194; Robert Siegfried. *From Elements to Atoms*. Op. cit., pp. 209–10.

78 This was a philosophical principle, not an empirical generalization, but it was nevertheless part of the consensus. So it was natural for Fourcroy to worry that he had to admit the existence of too many elements. See: Mi Gyung Kim. *Affinity*. Op. cit., p. 377.

79 Alistair Duncan. *Laws and Order*. Op. cit., p. 194.

The Affinity schema was probably more complex than this. In his list of "laws" of affinity, Fourcroy had statements that modified A-5", adding that affinity as a force works only through direct contact (not at a distance), and various additions to A-3 covering special cases of affinity, such as affinities through an intermediary substance, and reversible affinities.

80 Bernadette Bensaude-Vincent and Isabelle Stengers. *A History of Chemistry*. Op. cit., p. 70.

81 Ibid., p. 71.

Berthollet observed this phenomenon not only in industry but also in natural processes involving both large scale and a continuous removal of products, like the spontaneous production of a salt lake with banks covered with soda seen while accompanying the expedition to Egypt led by Napoleon in 1800.

82 Mi Gyung Kim. *Affinity*. Op. cit., p. 422.

83 William H. Brock. *The Chemical Tree*. Op. cit., p. 108.

84 Mi Gyung Kim. *Affinity*. Op. cit., pp. 425–6.

85 Ibid., pp. 333–7, 398–9.

The expression "a community of opinions" was Lavoisier's.

86 Ibid., pp. 398–9 (Berthollet) and p. 333 (Guyton).

The manner in which Kirwan was convinced to switch sides was more complex. See discussion of his case in: Philip Kitcher. *The Advancement of Science*. Op. cit., pp. 282–8.

87 William H. Brock. *The Chemical Tree*. Op. cit., p. 120.

88 John Murray. *A System of Chemistry*. Volumes I to IV (Edinburgh: Longman, Hurst & Rees, 1806).

89 Ibid., V. I, p. 144.

Murray acknowledges the precarious ontological status of caloric: is it a subtle, imponderable substance or a state of vibration of substances? But he argues that although we may not know its causes, we know its effects: to make substances expand and therefore to provide an explanation for the liquid and gas states.

90 Ibid., V. II, p. 124.

Joseph Black is credited with all the pioneering work on the ontology of fire, while two of his disciples, Adair Crawford and William Irvine, are given the credit for the concepts of specific heat and latent heat (pp. 118–20)

91 Bernadette Bensaude-Vincent. "The Chemists Balance for Fluids: Hydrometers and Their Multiple Identities, 1770–1810." In *Instruments and Experimentation in the History of Chemistry*. Op. cit., pp. 153–4.

The author argues that while the hydrometer was presented in 1751 as a phenomenon illustrating Archimedes' Law, by 1792 it had become an instrument for the identification of substances by their density.

92 John Murray. *A System of Chemistry*. Op. cit., V. II.

Murray uses traditional properties like smell or taste alongside specific gravity to determine the identity of oxygen (p. 17), nitrogen (p. 26), hydrogen (p. 33), as well as their binary compounds, such as ammonia (p. 74). In V. III, he uses it to identify metals (pp. 536–7).

93 Ursula Klein and Wolfgang Lefévre. *Materials in Eighteenth-Century Science*. Op. cit., p. 224.

Although the fact that medical virtues were lost if analysis was carried out too far was recognized by chemists like Boerhaave, it was Gabriel François Venel who first listed the proximate principles that possessed irreducible properties: essential oil, fatty oil, balsam, resin, gum, extracts, etc.

94 John Murray. *A System of Chemistry*. Op. cit.,V. I., p. 72.

To ensure that the reader understands the central importance of the statement that a compound's properties are not intermediate but novel, Murray refers to it as "the law of chemical attraction."

95 Bernadette Bensaude-Vincent and Isabelle Stengers. *A History of Chemistry*. Op. cit., pp. 72–4.

The authors argue that in the controversy between Berthollet and Proust over definite proportions, the metallic oxides that were at the center of the debate were not pure compounds but mixtures of oxides of different types, but the experimental techniques available to them could not tell them apart. Murray refers to this debate in: John Murray. *A System of Chemistry*. V. I. Op. cit., pp. 128–30.

96 Pierre Joseph Macquer. *Elements of the Theory and Practice of Chemistry*. Vol. I. Op. cit., pp. 54–5 (gold and silver) and pp. 93–94 (tartar).

97 John Murray. *A System of Chemistry*. V. II. Op. cit., p. 17 (oxygen), p. 25 (nitrogen), p. 31 (hydrogen), p. 83 (potassa), p. 281 (carbon).

98 Ibid., V. III.

Murray lists 28 metals, many of recent discovery, and asserts that they have not yet been decomposed and that there are no grounds to believe they will (p. 129). Oxides are discussed as the most important compounds that metals can form (p. 133).

99 Ibid., V. II.

Ammonia, which was listed by Lavoisier as an element, is treated here as a compound (p. 62). But potash and soda, which Humphry Davy would soon show to be compounds of the newly isolated potassium and sodium, are classified as not yet decomposed substances, that is, as elements (pp. 77, 90).

100 Ibid., V. II, p. 68 (statement 4), p. 404 (statement 5), p. 206 (statements 6 and 7), Vol I, pp. 110–14 (statement 8).

Statement 8 is, of course, a joint statement of the "laws" of definite proportions and of multiple proportions, attributed to Joseph Proust and John Dalton respectively. When this textbook was being written, the truth of this empirical generalization was being questioned in a debate between Proust and Berthollet. The debate fizzled out and these "laws" became consensus soon after that. It is therefore uncertain whether this statement should be included as part of the consensus of 1800, but it certainly was consensual by 1810.

101 Ibid., V. I, Notes, p. 84. In this case, the question is phrased explicitly as a Why question. But not everything that is considered problematic in this textbook is phrased that way. Other expressions include: "A satisfactory explanation has yet to be found for ...," "Whatsoever may be the cause of ...," "How is this to be reconciled with the theory that ..."

102 Ibid., V. II, p. 49.

103 Ibid., V. I, Notes, p. 15.

104 Ibid., V. III, pp. 159–60.

105 Ibid., V. I.

Among the changes to the Affinity schema there was the consolidation of statement A-5", asserting that outcomes are a result of a balance of forces (pp. 34–47), and the replacement of A-4 by A-4', denying that affinity forces were constant (pp. 81–5). The Part-to-Whole schema was also changed by the replacement of PW-2 by PW-2', asserting that compounds have properties different from those of its components (p. 72), and by changing PW-5 to PW-5', asserting that a compound's properties depend on the nature of its components as well on their relative proportions (V. II, p. 206).

106 Ibid., V. I, Notes, pp. 123–5.

107 Ibid., V. II, pp. 21–2.

The author lists the affinities of oxygen in the "dry way" (affected by heat), as given by Lavoisier, but he notes that it is useless to list these affinities in the "humid way" (affected by solvents), for these are too uncertain and complicated.

108 There were, to be sure, several linguistic issues involved, such as the choice of language (Latin or French); the choice of words; and the choice of suffixes to distinguish substances varying in the proportions of their components. The defense and promotion of the new nomenclature also touched on the metaphysics of language. Specifically, Lavoisier mobilized the ideas of Condillac about language as an analytical tool, and the need for perfecting it to place it at the service of thought. See: Maurice P. Crossland. *Historical Studies in the Language of Chemistry*. Op. cit., pp. 178–9.

109 This particular version of the compositional hierarchy was published later by Fourcroy. See: William H. Brock. *The Chemical Tree*. Op. cit., p. 115.

Our reference textbook not only acknowledges the importance of the new nomenclature, but its presentation is structured following a compositional hierarchy, discussing first the simple gases, then moving on to discuss binary combinations of these gases, and so on. See: John Murray. *A System of Chemistry*. V. II. Op. cit., pp. 155–6.

Chapter 2: Organic Chemistry

The specialization of cognitive tools

1 William H. Brock. *The Chemical Tree. A History of Chemistry* (New York: W.W. Norton, 2000), pp. 130–1.

Jeremias Richter is credited with the systematic work in the 1790s that led to these proportions. The credit for the tabulation of Richter's results in 1802 that led to the concept of equivalent weight goes to Ernst Fischer.

2 Gustavus Detlef Hinrichs. *The True Atomic Weights of the Chemical Elements and the Unity of Matter* (New York: B. Westermann, 1894), pp. 19–21.

3 John Dalton, who pictured atoms as spherical particles endowed with mechanical properties (size, form, position) and surrounded by an atmosphere of heat, can be said to have used atomic weights as units of composition. Jöns Jacob Berzelius, on the other hand, believed that all we needed was to think of small portions of a substance, without any particular scale or mechanical properties. See: Ursula Klein. *Experiments, Models, and Paper Tools* (Stanford, CA: Stanford University Press, 2003), p. 19.

The terms "units of composition" and "units of combination" are used to mark this distinction in: Bernadette Bensaude-Vincent and Isabelle Stengers. *A History of Chemistry* (Cambridge, MA: Harvard University Press, 1996), p. 114.

4 Ursula Klein. Op. cit., p. 23.

Klein argues that of the 13 organic compounds analyzed by Berzelius in 1815, only in eight cases was he able to transform the measurements into integers without adjusting them.

5 William H. Brock. Op. cit., pp. 143–5.

These empirical generalizations were known as the "law" of definite proportions and the "law" of multiple proportions. There was, in addition, a third empirical regularity: when gases combined with each other, they did so in simple proportions. For example, 100 volumes of nitrogen and 300 volumes of hydrogen yielded 200 volumes of ammonia. This other empirical generalization, due to Joseph Louis Gay-Lussac, was important because it served as a bridge between gravimetric and volumetric analyses (ibid., p. 163).

6 Bernadette Bensaude-Vincent and Isabelle Stengers. *A History of Chemistry*. Op. cit., p. 74.

7 Ursula Klein and Wolfgang Lefévre. *Materials in Eighteenth-Century Science. A Historical Ontology* (Cambridge, MA: MIT Press, 2007), p. 268.

The extension of stoichiometric methods to organic chemistry was performed by the chemists Jöns Jacob Berzelius and Michel Eugéne Chevreul between 1815 and 1823 (ibid., pp. 280–3).

8 Maurice. P. Crosland. *Historical Studies in the Language of Chemistry* (New York: Dover, 1978), pp. 265–9.

9 Ibid., pp. 271–7.

The formulas were first introduced by Berzelius in 1813. By 1814, he had given formulas to 50 different substances. Berzelius favored the use of superscripts for units of combining weights, and introduced several ways of abbreviating long formulas (such as using dots for oxygen), but neither of these conventions survived. The use of subscripts still in use today was proposed by Justus Von Leibig in 1834.

10 William H. Brock. *The Chemical Tree*. Op. cit., p. 212.

11 Mary Joe Nye. *From Chemical Philosophy to Theoretical Chemistry* (Berkeley, CA: University of California Press, 1993), pp. 85–6.

12 Ursula Klein. *Experiments, Models, and Paper Tools*. Op. cit., pp. 157–9.

The example given in the text uses the combinatorial units (and format for the radical) that became accepted after 1860, in which the number of units was halved. The original rational formula for the oil of bitter almonds, as written by Leibig and Friedrich Wöhler, was:

$$(14 C + 10 H + 2 O) + 2 H$$

13 Ibid., pp. 210–11.

The large variety of metamorphoses that a substance like alcohol could undergo, for example, allowed series of hundreds of substances to be derived from it. Alcohol was, in this sense, the "oxygen" of the 1820s: reactions using alcohol as starting point were compared by Liebig to the all-important combustion reactions of the eighteenth century. But alcohol would not have achieved this status without the spread of the practice of manipulating formulas on paper as a means to define radicals (ibid., p. 226).

14 Ibid., p. 196.

15 Bernadette Bensaude-Vincent and Isabelle Stengers. *A History of Chemistry*. Op. cit., p. 130.

The credit for establishing the possibility of substituting hydrogen by chlorine goes to Jean-Baptiste Dumas. But Dumas hesitated on the subject, perhaps out of fear to challenge Berzelius's electrochemical dualism, so it was his young assistant Auguste Laurent who first clearly expressed the concept of substitution.

16 William H. Brock. *The Chemical Tree*. Op. cit., p. 237.

17 Alan J. Rocke. *Image and Reality: Kekule, Kopp, and the Scientific Imagination* (Chicago: University of Chicago Press, 2010), p. 28.

One exception to this non-iconic interpretation of type formulas was Alexander Williamson, who believed the formulas were related to the structure of invisible molecules like an orrery to a planetary system.

18 Marya Novitski. *Auguste Laurent and the Prehistory of Valence* (Chur: Harwood Academic Publishers, 1992), p. 110.

19 Colin A. Russell. *The History of Valency* (New York: Humanities Press, 1971), p. 45.

In addition to Dumas, the two chemists who developed the type approach to classification were Laurent and his partner Charles Gerhardt. The former was ontologically committed to both atoms and molecules, introducing the distinction for the first time, while the latter remained uncommitted to invisible entities until the end of his life.

20 Bernadette Bensaude-Vincent and Isabelle Stengers. *A History of Chemistry*. Op. cit., p. 111.

21 Marya Novitski. *Auguste Laurent and the Prehistory of Valence*. Op. cit., pp. 36–7, 42–6.

The phenomenon of isomorphism was introduced into organic chemistry by Laurent, who used his crystallographic background to propose a spatial model of substitutions.

22 William H. Brock. *The Chemical Tree*. Op. cit., p. 225.

23 Ursula Klein. *Experiments, Models, and Paper Tools* (Stanford, CA: Stanford University Press, 2003), pp. 88–9.

The original research on this reaction was performed by Fourcroy and Louis Nicolas Vauquelin in 1797. By the 1830s it was common for chemists like Leibig to think of this reaction as a dehydration.

24 Ibid., pp. 120–8.

The chemists who created this model in 1827 were Dumas and Polydore Boullay.

25 William H. Brock. *The Chemical Tree*. Op. cit., p. 214.

Leibig and Wöhler stumbled upon the phenomenon of isomerism when studying silver cyanate and silver fulminate in 1823. Because these substances had identical composition but different properties, each of them assumed that the other one had made an analytical mistake. But then Wöhler showed in 1828 that urea (extracted from animal urine) had the same composition as ammonium cyanate, proving that the problem was real. Finally, Berzelius named the phenomenon as "isomerism" in 1830, when failing to detect any compositional difference between racemic and tartaric acids.

26 Peter J. Ramberg. *Chemical Structure, Spatial Arrangement: The Early History of Stereochemistry 1874–1914* (Aldershot: Ashgate, 2003), pp. 51–2.

27 Ibid., p. 45.

The formulas for lactic acid's isomers were created by the Scottish chemist Alexander Crum Brown. He originally enclosed the Berzelian symbols within circles, but these were eventually dropped.

28 Ibid., pp. 57–63.

The chemist who introduced three-dimensional structural formulas in 1874 was Jacobus van't Hoff. Joseph Le Bel reached similar conclusions independently in the same year.

29 Marya Novitski. *Auguste Laurent and the Prehistory of Valence.* Op. cit., pp. 101–2.

30 Colin A. Russell. *The History of Valency* (New York: Humanities Press, 1971), p. 42.

The chemist who published this observation in 1852 and who coined a term for this combining capacity ("atomicity") was Edward Frankland.

31 Marya Novitski. *Auguste Laurent and the Prehistory of Valence.* Op. cit., pp. 114–16.

The credit for recognizing the universality of the numerical regularities goes to William Odling, an advocate of the type approach. Odling was an anti-atomist for whom the concept of hydrogen-replacing power was just a useful numerical value, not related to either affinity or to a binding force exerted by actual microscopic particles. The term "affinity unit" was coined by August Kekulé to link hydrogen-replacing power to the old concept of a disposition to combine.

32 Colin A. Russell. *The History of Valency.* Op. cit., pp. 110–25.

The idea of a maximum combining power (saturation capacity) was due to Frankland. The idea of the polyvalency (multiple combining capacity) of radicals was Williamson's, while that of elements was Adolph Wurtz's. Credit for the tetravalency of carbon is shared by Odling and Kekulé. And the self-linking of carbon atoms to form chains or rings was the idea of both Kekulé and the Scottish chemist Archibald Scott Couper.

33 Ibid., pp. 242–3.

34 This way of counting affinity units, or of calculating the number of atomicities, appears in the textbook by Wurtz which will be used as reference for the period 1800–1860 in the next section of this chapter. See: Adolphe C. Wurtz. *An Introduction to Chemical Philosophy According to the Modern Theories* (London: J.H. Dutton, 1867), pp. 136–8.

35 Colin A. Russell. *The History of Valency*. Op. cit., pp. 198–9.

By the end of the nineteenth century, phosphorous, sulphur, and oxygen were known to display variable valency. Carbon had to wait until 1904 for its valency number to be acknowledged as variable.

36 Peter J. Ramberg. *Chemical Structure, Spatial Arrangement*. Op. cit., p. 281.

37 Ibid., pp. 288–9.

This other model of affinity was created by the German chemist Alfred Werner in 1893.

38 William H. Brock. *The Chemical Tree*. Op. cit.

There were debates in the first half of the century between Berzelius and Liebig as defenders of the radical approach, and Dumas, Laurent, and Gerhardt who stood by the type approach (with the added twist of a priority dispute between Dumas and Laurent) (p. 226). Then there was the debate in the second half between defenders of the structural approach to formulas, like van't Hoff, and those who attacked them, like the very vitriolic Hermann Kolbe (p. 268). And finally, towards the end of the century, the debate between those who could not conceive of a property like valency being variable, like Kekulé, and those who knew that phenomena exhibiting variability would not go away, like Werner (pp. 575–6).

39 Mary Jo Nye. *From Chemical Philosophy to Theoretical Chemistry: Dynamics of Matter and Dynamics of Disciplines* (Berkeley, CA: University of California Press, 1993), p. 143.

The perfect example of an authoritative hold-out was the French chemist Marcelin Berthelot, who until 1890 continued to use equivalent units, writing the formula for water as HO, and who never accepted the legitimacy of structural formulas.

40 William H. Brock. *The Chemical Tree*. Op. cit., pp. 242–3.

Hence, it was beneficial that advocates of the radical approach did not switch sides, even after types had been declared winners, since that kept the problem of variable valency alive.

41 Ibid., p. 621.

42 Marya Novitski. *Auguste Laurent and the Prehistory of Valence*. Op. cit., pp. 12–17.

43 Eric R. Scerri. *The Periodic Table: Its Story and Its Significance* (Oxford: Oxford University Press, 2007), pp. 126–30.

44 Ibid., p. 184.

45 William H. Brock. *The Chemical Tree*. Op. cit., p. 468.

The first polar model of the bond was created by the physicist J. J. Thomson.

46 Ibid., pp. 472–3.

The first non-polar model of the bond was created by the American chemist Gilbert Lewis in 1916.

47 Ibid., pp. 592–3.

The synthesis of the first 25 years of work on the chemical bond was performed by the chemist Nevil Sidwick.

48 Eric R. Scerri. *The Periodic Table*. Op. cit., p. 209.

49 Ibid., pp. 155–6.

50 Ibid., pp. 205–11.

Lewis was the creator of the cubic model, but his sketches of the model dating to 1902 were not immediately published.

51 Ibid., pp. 188–92.

In 1913, the Danish physicist Ernst Bohr pioneered the method of building up atoms layer by layer, using as his guides valency numbers and the Periodic Table. Bohr's modeling approach was also guided by the spectroscopic signatures of elements, a phenomenon that will be discussed in the next chapter.

52 Ibid., p. 212.

In 1916, partly influenced by Bohr, Lewis created the concept of an outer shell populated by "valence electrons," a shell that in the case of inert gases contained exactly eight electrons. But Lewis worked out his models only up to element 35. The next step, building up models up to element 92, was taken by another American chemist, Irving Langmuir, in 1919.

53 Ian Hacking. *Representing and Intervening* (Cambridge: Cambridge University Press, 1983), p. 146.

Hacking argues that what we consider real varies depending on whether we think of reality as that which we can correctly represent, or as that which we can affect and which can affect us. Spraying electrons on screens falls into the latter category.

54 Ibid., pp. 263–5.

From personal to consensus practice 1800–1900

1 Ursula Klein and Wolfgang Lefévre. *Materials in Eighteenth-Century Science. A Historical Ontology* (Cambridge, MA: MIT Press, 2007), p. 248.

2 Ibid., p. 249.

3 Ibid., pp. 251–2.

4 Ibid., p. 253.

Berzelius was a vitalist, but of the materialist kind. He believed that the synthesis of organic compounds involved a special affinity present only in living things.

5 William H. Brock. *The Chemical Tree. A History of Chemistry* (New York: W. W. Norton, 2000), p. 214.

6 Ibid., pp. 150–4.

Humphrey Davy had observed this phenomenon and made the connection to affinity earlier than Berzelius, but it was the latter who developed this insight into an electrochemical model of composition and reactivity.

7 Ibid., p. 212.

8 Ibid., p.163.

9 Ursula Klein. *Experiments, Models, and Paper Tools. Cultures of Organic Chemistry in the Nineteenth Century* (Stanford, CA: Stanford University Press, 2003), pp. 12–13.

Gay-Lussac assumed that the ratios between volumes had to be a small number (1, 2, or 3), an arbitrary assumption that Berzelius discarded when he adopted his results.

10 Ibid., p. 113.

11 Ursula Klein and Wolfgang Lefévre. *Materials in Eighteenth-Century Science. Op. cit.*, p. 265.

In 1833, Berzelius finally accepted that substances of organic origin like alcohol could be composed of inorganic ones, but he never classified carbon-based substances created in the laboratory as organic.

12 Alan J. Rocke. "Organic Analysis in Comparative Perspective: Liebig, Dumas, and Berzelius, 1811–1837." In *Instruments and Experimentation in the History of Chemistry*. Edited by Frederic L. Holmes and Trevor H. Levere (Cambridge, MA: MIT Press, 2000), p. 274.

13 Ibid., p. 275.

14 Ibid., pp. 282–4.

15 Maurice P. Crossland. *Historical Studies in the Language of Chemistry* (New York: Dover, 1978), p. 304.

16 William H. Brock. *The Chemical Tree*. Op. cit., p. 212.

In 1833, Liebig and Robert Kane called attention to the role of conventions with regard to a component of ethyl chloride, C_4H_8, a group that remained invariant in some reactions, but that could also be formulated as C_4H_{10} on the basis of other transformations.

17 Ursula Klein. *Experiments, Models, and Paper Tools*. Op. cit., pp. 189–92.

Because the manipulation of formulas was crucial to compensate for the limitations of analysis, Klein refers to them as "paper tools," playing the same enabling role as laboratory instruments.

18 Ibid., p. 195.

Dumas found these results compelling enough to change his position on the binary constitution of organic substances, and on the unchanging nature of radicals, despite the fact that this went against his professional interests and against the interests of most members of the chemical community (ibid., pp. 198–9).

19 Ibid., pp. 115–16.

This difficulty had become evident since 1820 when Gay-Lussac showed that the explanation for the production of ether by the dehydration of alcohol did not take into account several byproducts, such as sulfovinic acid. Dumas (working with Polydore Boullay) created the first full model of this important reaction in 1827, modeling the transformation as *two parallel reactions*, one in which ether was produced by dehydration, and another in which sulfovinic acid was simultaneously created from successive degradations. He arrived at this hypothesis not through the balancing of masses measured in the laboratory but by balancing the combinatorial units in the formulas (ibid., pp. 120–8).

20 Marya Novitski. *Auguste Laurent and the Prehistory of Valence* (Chur: Harwood Academic Publishers, 1992), pp. 32–3.

21 Ursula Klein. *Experiments, Models, and Paper Tools*. Op. cit., pp. 196–7.

22 Marya Novitski. *Auguste Laurent*. Op. cit., pp. 36–45.

23 William H. Brock. *The Chemical Tree*. Op. cit., p. 224.

24 Marya Novitski. *Auguste Laurent*. Op. cit., pp. 76–8.

25 William H. Brock. *The Chemical Tree*. Op. cit., pp. 216–18.

26 Marya Novitski. *Auguste Laurent*. Op. cit., p. 108.

27 Bernadette Bensaude-Vincent and Isabelle Stengers. *A History of Chemistry* (Cambridge, MA: Harvard University Press, 1996), p. 135.

28 William H. Brock. *The Chemical Tree*. Op. cit., p. 228.

29 Bernadette Bensaude-Vincent and Isabelle Stengers. *A History of Chemistry*. Op. cit., pp. 136–7.

30 Alan J. Rocke. *Image and Reality: Kekule, Kopp, and the Scientific Imagination* (Chicago: University of Chicago Press, 2010), p. 19.

31 Ibid., p. 20.

32 William H. Brock. *The Chemical Tree*. Op. cit., p. 235.

Williamson went on to show that inorganic salts and acids, as well as other organic compounds, also belonged to the water type.

33 Ibid., p. 28.

Williamson's realism about molecular structure led him to propose dynamic reaction mechanisms in which elementary atoms and groups of atoms were detached from a molecule and moved to another in a continuous way. He hypothesized that in the original ether-producing reaction, ethyl radicals (C_2H_5) moved continuously from alcohol to sulphuric acid and then back to alcohol. Once equilibrium was reached, the alcohol molecules that ended up without this radical became water; those that had acquired a radical became ether; while the molecules of sulphuric acid that had acquired an ethyl radical became one of the important byproducts of the reaction, sulfovinic acid. Kinetic motion and the concept of a natural termination point for a reaction defined by the stability of the final products were for the first time introduced into a model of a transformation (ibid., p. 21).

34 Marya Novitski. *Auguste Laurent*. Op. cit., p. 113.

35 Colin A. Russell. *The History of Valency* (New York: Humanities Press, 1971), pp. 81–6.

36 Marya Novitski. *Auguste Laurent*. Op. cit., p. 116.

37 Colin A. Russell. *The History of Valency.* Op. cit., p. 120.

38 Ibid., pp. 35–8.

39 Ibid., p. 42.

40 Alan J. Rocke. "The Quiet Revolution of the 1850s." In *Chemical Science in the Modern World.* Edited by Seymour H. Mauskopf (Philadelphia, PA: University of Pennsylvania Press, 1993), pp. 94–5.

The author uses the Kuhnian term "conversion" to characterize the switching of sides by Adolphe Wurtz (ibid., p 101), and of the last hold-out and most bitter opponent, Adolphe Kolbe (ibid., p. 105). Some of the actors themselves used this word. But it is clear that Rocke does not believe this was a Gestalt-switch between two monolithic paradigms: every participant had residual doubts about one or another item in dispute, and there was as much continuity as discontinuity.

41 Mary Joe Nye. *Introduction. The Question of the Atom: from the Karlsruhe Congress to the First Solvay Conference, 1860–1911* (Los Angeles: Tomash Publishers, 1986), pp. xiii–xiv.

It was an Italian chemist, Stanislao Cannizzaro, who actively promoted the new standards (uniform atomic weights, the diatomic nature of some gases, Avogadro's hypothesis) as necessary for didactic purposes. Cannizzaro also defended the identity of the "chemical atom" (the limit of chemical analysis) and the "physical atom" (units of composition), but this identity remained highly controversial until 1911.

42 William H. Brock. *The Chemical Tree.* Op. cit., pp. 242–3.

Berzelius, the father of electrochemical dualism, responded to the challenge posed by the "impossible" substitution of hydrogen by chlorine by adding electrically neutral "copula" to the radical schema. Kolbe, his German disciple, and last hold-out on the radical side, usefully extended these ideas as the controversy unfolded, while his ally Frankland used the same extension as inspiration for an early form of the concept of valency. Thus, the "losers" had plenty to contribute to the new consensus. On the other side, advocates of the type approach like Gerhardt correctly saw that some elements of electrochemical dualism needed to be incorporated into their own framework, rejecting only the concept that all substances had a binary constitution (ibid., p. 240).

43 Adolphe C. Wurtz. *An Introduction to Chemical Philosophy According to the Modern Theories* (London: J.H. Dutton, 1867).

This text, published in 1864, is not really a textbook but an introductory course. Yet, it will serve the purpose of indicating what had become accepted in the chemical community by 1860, Wurtz having attended the conference and therefore being in a position to summarize the partial consensus forged there. On the other hand, Wurtz was a convinced atomist, a stance that was certainly not consensual at this time, so his ontological commitments will be played down.

44 John Murray. *A System of Chemistry*. Volume IV (Edinburgh: Longman, Hurst & Rees, 1806), pp. 2–4.

45 Adolphe C. Wurtz. *An Introduction to Chemical Philosophy*. Op. cit., pp. 145–6.

In these pages Wurtz tries to figure out why other elements having the same combinatorial capacity as carbon, such as silicon, do not produce as many compounds. Unlike the carbides of hydrogen, he says, silicides of hydrogen are spontaneously inflammable in air, hence, unstable.

46 Ibid., p. 106.

The idea of a homologous series, and that of CH2 as a module for some series, was put forward by Gerhardt between 1842 and 1846. See: William H. Brock. *The Chemical Tree*. Op. cit., p. 231.

47 John Murray. *A System of Chemistry*. V. I. Op. cit., pp. 110–14.

48 Adolphe C. Wurtz. *An Introduction to Chemical Philosophy*. Op. cit., p. 9.

This is, of course, a joint statement of the "laws" of definite proportions and of multiple proportions, attributed to Joseph Proust and John Dalton respectively.

49 Ibid., pp. 11–12.

This general statement is known as the "law" of Gay-Lussac, an empirical generalization from many independently confirmed particular statements like:

Two volumes of hydrogen combine with one volume of oxygen to form two volumes of aqueous vapor.

Three volumes of hydrogen combine with one volume of nitrogen to form two volumes of ammoniacal gas.

50 Ibid., pp. 21–2.

51 Ibid., pp. 20–1.

This general statement is known as the "law" of Dulong and Petit, an empirical generalization from many particular statements like:

Multiplying the specific heat of sulphur (0.1880) by its atomic weight (201.15) equals .379.

Multiplying the specific heat of lead (0.0293) by its atomic weight (1294.5) equals .3793.

52 This process of mutual adjustment has been discussed under the rubric of a "dialectic of resistance and accommodation" by Andrew Pickering. See: Andrew Pickering. *The Mangle of Practice. Time, Agency, and Science* (Chicago: University of Chicago Press, 1995), pp. 22–3.

53 Bernadette Bensaude-Vincent and Isabelle Stengers. *A History of Chemistry*. Op. cit., pp. 122–3.

These exceptions confronted chemists with problems like these:

Why do phosphorous and arsenic have twice the vapor density they should have given their atomic weights?

Why do mercury and cadmium have half the vapor density they should have given their atomic weights?

Adolphe C. Wurtz. *An Introduction to Chemical Philosophy*. Op. cit., pp. 47–9.

54 Ibid., p. 33.

55 Ibid., p. 49.

Statement 5 is known as "Avogadro's hypothesis," but it is also attributed here to Berzelius and to the physicist André-Marie Ampére.

56 Ibid., p. 121.

Wurtz uses the terms "atomicity" and "polyatomicity" rather than "valency" and "polyvalency."

57 Ibid., p. 118.

58 Ibid., p. 106.

59 Ibid., p. 59.

60 Ibid., p. 111.

61 Ibid., p. 41.

62 Ibid., p. 96.

Wurtz refers to type formulas as instruments of explanation and classification.

63 Ibid., pp. 90–2.

64 Ibid., p. 99.

65 Ibid., p. 117.

66 Mary Joe Nye. *Introduction. The Question of the Atom*. Op. cit., pp. xiii–xiv.

67 Alan J. Rocke. *Image and Reality*. Op. cit., pp. 191–2.

68 Peter J. Ramberg. *Chemical Structure, Spatial Arrangement: The Early History of Stereochemistry 1874–1914* (Aldershot: Ashgate, 2003), pp. 31–4.

The existence of optical isomers of tartaric acid was established by Louis Pasteur in 1847. The phenomenon of rotatory polarization itself was first observed by Jean-Baptiste Biot in 1815, who went on to refine the polarimeter and establish standards for calculating rotation angles in the 1830s.

69 Colin A. Russell. *The History of Valency*. Op. cit., pp. 68–71.

70 Ibid., pp. 242–3.

71 Alan J. Rocke. *Image and Reality*. Op. cit., pp. 201–5.

72 Peter J. Ramberg. *Chemical Structure, Spatial Arrangement*. Op. cit., p. 43.

73 Ibid., p. 48.

Wislicenus introduced in 1873 a classification of the different kinds of isomerism: positional isomers were those in which the position of a radical (like hydroxyl) differed between the variants; core isomers were those that differed in their carbon skeleton; and finally, geometric isomers were those that differed in the spatial arrangement of their components

74 Maurice P. Crossland. *Historical Studies in the Language of Chemistry*. Op. cit., pp. 331–4.

75 William H. Brock. *The Chemical Tree*. Op. cit., pp. 254–5.

76 Christoph Meinel. "Molecules and Croquet Balls." In *Models. The Third Dimension of Science*. Edited by Soraya de Chadarevian and Nick Hopwood (Stanford, CA: Stanford University Press, 2004), pp. 249–52.

The ball-and-stick models were created by August Wilhelm Hoffmann.

77 Peter J. Ramberg. *Chemical Structure, Spatial Arrangement*. Op. cit., pp. 324–5.

78 Alan J. Rocke. *Image and Reality*. Op. cit., p. 241.

Van't Hoff was aware that to see the validity of his argument other chemists would have to be able to perform the mirror-image transformation in their heads, and since he could not count on his colleagues to have this skill, he created cardboard tetrahedrons that could be manipulated to this end, mailing them to all contemporary chemists working on questions of molecular structure.

79 Peter J. Ramberg. *Chemical Structure, Spatial Arrangement*. Op. cit., pp. 74–6.

In 1875, Gustav Bremer tested van't Hoff's model in the laboratory, using malic acid as his model substance. The model predicted that in addition to the naturally occurring isomer (an ammonium salt of malic acid) that rotated the plane of vibration of light to the left, there should be a mirror-image variant that rotated it to the right. Bremer was able to synthesize this artificial variant.

80 Ibid., pp. 70–1.

The statement that possession of an asymmetric carbon determines optical activity "saved the phenomena," as positivists are fond of saying. But van't Hoff did offer a speculative causal explanation: tetrahedral carbon atoms joined by vertices or edges formed a helical arrangement that could twist passing light.

81 Ibid., pp. 101–4.

82 Bernadette Bensaude-Vincent and Isabelle Stengers. *A History of Chemistry*. Op. cit., p. 144.

83 Ibid., p. 153.

84 Herbert Morawetz. *Polymers. The Origins and Growth of a Science* (New York: Dover, 1985), p. 18.

85 Ibid., p. 20.

86 Peter J. Ramberg. *Chemical Structure, Spatial Arrangement*. Op. cit., p. 69.

87 Ibid., p. 264.

88 Ibid., p. 271.

89 Ibid., pp. 268–71.

90 August Bernthsen. *A Textbook of Organic Chemistry* (New York: D. Van Nostrand, 1912).

The index of this translation, and the chapters on open-chain carbon compounds and cyclic carbon compounds, are almost identical with the 1902 German edition. These will be the sections from which most of the examples will be taken, so it can be assumed that the textbook provides a good point of reference for the content of consensus practice in 1900.

91 Ibid., pp. 16–17.

92 Ibid., pp. 87, 138.

93 Ibid., pp. 13, 44.

94 Ibid., pp. 227–8.

95 Ibid., pp. 32, 322.

96 Ibid., p. 244.

97 Ibid., p. 323.

98 Ibid., p. 38.

99 Ibid., p. 39.

100 Ibid., p. 66.

101 Ibid., pp. 30, 42, 49, 54, 66, 82, 86, 121, 140.

102 Ibid., p. 56.

103 Ibid., p. 16.

104 Ibid., pp. 90–1.

105 J. Erik Jorpes. *Jacobus Berzelius: His Life and Work* (Berkeley, CA: University of California Press, 1970.), p. 112.

106 August Bernthsen. *A Textbook of Organic Chemistry*. Op. cit., pp. 666–7.

The isolation of "press juice" from yeast cells and the proof that it could act as ferment were achieved by Hans Buchner. It took at least five more years for this to be generally accepted given that the contrary thesis, that living organisms were involved, was maintained by the authority of Louis Pasteur. On the controversy and its slow resolution see: David Dressler and Huntington Potter. *Discovering Enzymes* (New York: Scientific American Press, 1991), pp. 47–51.

107 Ibid., p. 669.

108 Ibid., p. 126.

109 Ibid., p. 245.

Chapter 3: Physical Chemistry

The hybridization of cognitive tools

1 Frederick L. Holmes. *From Elective Affinities to Chemical Equilibria: Berthollet's Law of Mass Action*. Chymia. Vol. 8 (Berkeley, CA: University of California Press, 1962), pp. 106–13.

The first model was created by Torbern Bergman, the second by Claude-Louis Berthollet.

2 Ibid., pp. 117–118.

3 Ibid., pp. 120–7.

Holmes argues that the neglect of Berthollet's model was only partial. Chemists like Joseph Louis Gay Lussac and Jöns Jacob Berzelius did try to modify the model to make it compatible with the concept of definite proportions, and attempted to produce experimental evidence for incomplete reactions. The time spent on these matters, however, cannot be compared to the amount of effort these chemists put into organic chemistry. Thus, in the end, the neglect may have been caused by a lack of human resources: most chemists were focused on compositional problems and did not have the time to tackle reactions.

4 Virginia M. Schelar. *Thermochemistry and the Third Law of Thermodynamics*. Chymia. Vol. 11 (Berkeley, CA: University of California Press, 1966), pp. 100–2.

The measurements were carried out by Germain Henri Hess between 1839 and 1842, a German chemist who also stated the first regularity about heats of reaction (the independence of the heat released from the reaction pathway) and coined the term "thermochemistry."

5 Ibid., pp. 108–10.

Over 3,500 measurements were performed by the Danish chemist H. P. Julius Thomsen starting in the 1850s, the results of his life work published between 1882 and 1886.

6 Trevor H. Levere. *Affinity and Matter. Elements of Chemical Philosophy, 1800–1865* (Oxford: Clarendon, 1971), pp. 89–91.

The English chemist Michael Faraday, working in the 1830s, deciphered the chemical mechanism of electrolysis.

7 Ibid., p. 87.

Faraday conducted the experiments that yielded these quantitative statements and stated the regular dependence between quantities of substances separating at the electrodes and their equivalent weights.

8 William H. Brock. *The Chemical Tree. A History of Chemistry* (New York: W.W. Norton, 2000), p. 375.

The German chemist Wilhelm Hittorf performed the meticulous experiments that yielded these quantitative statements as well as the general statement of the regularity.

9 Ibid., p. 376.

The measurement of molecular conductivities, and the statement of the regular dependence of conductivity and dilution, were the result of the work of the German physicist Friedrich Kohlrausch, performed between 1868 and 1875.

10 Ibid., pp. 363–4.

The measurements that established this problematic regularity were made by François Marie Raoult in the 1880s.

11 Wesley C. Salmon. *Scientific Explanation and the Causal Structure of the World* (Princeton, NJ: Princeton University Press, 1984), p. 136.

Salmon conceives of scientific explanations as involving two steps: an assembly of facts that are statistically relevant to whatever is to be explained (ibid., p. 45); and second, a causal account of the correlations between those facts, with causality defined in terms of production (a cause is an event that produces another event) and propagation (capacity to transmit causal influence) (ibid., p. 139). In this characterization, the ideal gas law supplies the first step, while a statistical mechanical model of collisions between particles provides the second step (ibid., p. 227).

12 Edwin C. Kemble. *Physical Science. Its Structure and Development* (Cambridge, MA: MIT Press, 1966), p. 381.

13 John W. Servos. *Physical Chemistry from Ostwald to Pauling. The Making of a Science in America* (Princeton, NJ: Princeton University Press, 1990), p. 41.

14 Edwin C. Kemble. *Physical Science*. Op. cit., pp. 382–4.

The first regularity in the behavior of ideal gases was discovered by Boyle in 1662. The second one was discovered by Gay-Lussac in 1802.

15 M. J. Klein. "The Physics of J. Willard Gibbs in His Time." In *Proceedings of the Gibbs Symposium*. Edited by D. G. Caldi and G. D. Mostow (Providence, RI: American Mathematical Society, 1990), p. 5.

The American mathematical physicist Willard Gibbs created an equation of state including variables for all these properties in 1873. Unfortunately, his abstract style and unwillingness to discuss particular applications prevented his model from influencing physical chemists at the time (ibid., p. 10).

16 Edwin C. Kemble. *Physical Science*. Op. cit., pp. 414–15.

17 Bernadette Bensaude-Vincent and Isabelle Stengers. *A History of Chemistry*. Op. cit., pp. 223–4.

18 John W. Servos. *Physical Chemistry from Ostwald to Pauling*. Op. cit., p. 32.

Jacobus van't Hoff, a Dutch chemist whose seminal work on organic chemistry was discussed in the previous chapter, created models for dilute solutions that were analogs of the models of Boyle and Gay-Lussac for ideal gases.

19 Ibid., p. 31.

The relevant experiments were carried out by the German botanist Friedrich Pfeiffer, who also introduced artificial membranes capable of withstanding greater pressures.

20 Alexander Findlay. *Osmotic Pressure* (London: Longmans, Green and Co., 1919), pp. 7–8.

21 Alexander Kipnis. "Early Chemical Thermodynamics: Its Duality Embodied in Van't Hoff and Gibbs." In *Van't Hoff and the Emergence of Chemical Thermodynamics*. Edited by Willem J. Hornix and S. H. W. M. Mannaerts (Delft: Delft University Press, 2001), p. 6.

Using an analog of the ideal engine to derive the thermodynamic equations was not, in fact, necessary. At the time van't Hoff was using the method of quasi-static cycles (that kept entropy constant) Gibbs had already created better equations, derived purely mathematically, that made the entropy not only constant but also a maximum at equilibrium, a more elegant and general result. But as indicated in a footnote above, Gibbs's work failed to be noticed by chemists.

22 Mary Jo Nye. *Before Big Science. The Pursuit of Modern Chemistry and Physics, 1800–1940* (New York: Twayne Publishers,1996), pp. 91–2.

The first "law" of thermodynamics, as well as the concept and measurement of the mechanical equivalent of heat, is attributed to James Joule. Joule was a student of John Dalton, and was influenced by Humphry Davy and Michael Faraday. He arrived at both the statement and the concept in 1837 while studying, among other things, the workings of an electrical motor.

23 Ibid., pp. 96–7.

The second "law" of thermodynamics is credited to Rudolph Clausius, who arrived at it in 1850 while trying to reconcile Joule's experimental results with the much earlier efforts of Sadi Carnot to explain the workings of the steam engine. The original version of the statement was not in terms of high- and low-quality energy (or fresh and exhausted gradients) but of the spontaneous flow from hotter to colder bodies. However, that heat flows from higher to lower temperatures was already known to chemists since the work of Joseph Black, so Clausius' contribution was to show that this flow was *irreversible* because energy becomes degraded as mechanical work is transformed into heat. He invented the concept of entropy to measure the degree of degradation.

24 Dilip Kondepudi. *Introduction to Modern Thermodynamics* (Chichester: John Wiley, 2007), p. 141.

Gibbs gave us the term "free energy," while chemists used "work of affinity," "thermodynamic potential," and "chemical energy." The term "chemical potential" is the one that ended up prevailing in the twentieth century.

25 Michael E. Fisher. "Phases and Phase Diagrams: Gibb's Legacy Today." In *Proceedings of the Gibbs Symposium*. Op. cit., pp. 41–8.

Gibbs introduced these diagrams into thermodynamics in 1873. The original diagram was three-dimensional with the equilibria distributed in a convex surface. The two-dimensional versions used by chemists were projections of this fundamental surface.

26 Wilhelm Ostwald. *The Fundamental Principles of Chemistry* (New York: Longmans, Green, 1909), p. 309.

27 Ibid., pp. 216–17.

The reason for the correspondence of geometric features and physical states is that the properties of compounds change by jumps (hence the kink in the curve) while those of mixtures change continuously (ibid., p. 119).

28 Ibid., pp. 167–8.

29 Christa Jungnickel and Russell McCormmach. *Intellectual Mastery of Nature. Theoretical Physics from Ohm to Einstein*. Vol. 1 (Chicago: University of Chicago Press, 1986), pp. 194–5.

30 Edwin C. Kemble. *Physical Science*. Op. cit., pp. 451–2.

The kinetic model of partial vaporization in a closed vessel was created by Clausius.

31 John W. Servos. *Physical Chemistry from Ostwald to Pauling*. Op. cit., pp. 28–9.

Van't Hoff showed how to derive the chemical analog of Clausius' microscopic model of equilibrium. He used as his starting point the work of two Danish chemists, Peter Waage and Cato Guldberg, who developed the first mathematical version of Berthollet's model of chemical reactions (the "law" of mass action) starting in 1864.

32 Wilhelm Ostwald. *The Fundamental Principles of Chemistry*. Op. cit., pp. 6–7.

Ostwald went as far as asserting that all properties are definable in terms of energy, or correspond to forms of energy, in the way in which weight corresponds to gravitational energy.

33 John W. Servos. *Physical Chemistry from Ostwald to Pauling*. Op. cit., pp. 34–5.

Staring in 1884, the Swedish chemist Svante Arrhenius performed the needed measurements of conductivity and discovered the regular variation of this property with the degree of dilution. The statement that dissolved substances exist partially dissociated in all solutions was first put forward by Alexander Williamson (whose work was described in the previous chapter) and adapted to electrolytic solutions by Clausius.

34 Ibid., pp. 36–7.

35 Riki G. A. Dolby. *Debates Over the Theory of Solution: A Study of Dissent in Physical Chemistry in the English-Speaking World in the Late Nineteenth and Early Twentieth Centuries*. Historical Studies in the Physical Sciences. Vol. 7 (Berkeley, CA: University of California Press, 1976), pp. 354–6.

There was an intense controversy at the time, starting in 1887, when physical chemists began publishing their own journal and brought international attention to the main ideas. It raged in Britain through the 1890s, and continued in the United States in the first decade of the twentieth century.

36 John. S. Rowlinson. "Introduction." In J. D. van Der Waals. *On the Continuity of the Gaseous and Liquid States* (New York: Dover, 2004), p. 10.

The simple changes proved inadequate but were used as a point of departure for a better model by the Dutch physicist Johannes Diderick van der Waals. In 1873 he extended the equation of state of ideal gases to the case particles with volume interacting in pairs.

37 Ibid., p. 21.

The idea of using a power series to capture the effects of higher densities on pressure was van der Waals', but others developed alternative power-law expansions, culminating in 1901 with the so-called virial expansion still in use today.

38 Alexander Findlay. *Osmotic Pressure*. Op. cit., p. 46.

A different way of increasing the cognitive value of mathematics in chemistry was to devise new ideal phenomena around which other families of equations could be generated. A chemist studying a real solution was interested in the number of substances in the mixture;

the state of each one of them (associated or dissociated); the change of volume that took place as a result of their mixing; the heat of dilution; the compressibility of the solution; and several other properties. Most of these characteristics were too complex to model mathematically. So instead, chemists could imagine an *ideal solution* in which only two substances entered into a mix without associating or dissociating; there were no interactions between the particles; the mixing of the two substances caused no changes in volume or heat; and if the solvent underwent any vaporization this vapor behaved like an ideal gas. This idealization creates a *tractable* phenomenon whose behavior can be fruitfully explored. An equation of state for an ideal solution can be created by representing the amount of solvent by N, and the amount of solute by n, and expressing their relation as $(n / N + n)$, that is, as the ratio of the number of particles of solute to the total number of particles (ibid., p. 55). If for convenience we express this ratio by the variable x, and expand it as a power series, then an equation of state for the ideal solution can be written like this:

$$PV = RT \ (x + 1/2 \ x^2 + 1/3 \ x^3 + 1/4 \ x^4 \ ...)$$

39 Dilip Kondepudi. *Introduction to Modern Thermodynamics*. Op. cit., p. 141.

The extension of chemical thermodynamics to near equilibrium states was performed in the 1920s by the Belgian physicist and mathematician Théophile DeDonder. The Belgian chemist Ilya Prigogine performed the extension to far from equilibrium states in the 1960s.

40 Ibid., p. 327.

41 Ibid., pp. 328–9.

42 David Jou, José Casas-Vázquez, and Georgy Lebon. *Extended Irreversible Thermodynamics* (Dordrecht: Springer, 2010), pp. 23–5.

43 David Jou, José Casas-Vázquez, and Manuel Criado-Sancho. *Thermodynamics of Fluids Under Flow* (Dordrecht: Springer, 2011). pp. 2–3.

44 David Jou, José Casas-Vázquez, and Georgy Lebon. *Extended Irreversible Thermodynamics*. Op. cit., p. 30.

One indication that ideal phenomena away from equilibrium are not just an extension of the old ideal phenomena is that properties like temperature and entropy are not well defined. The reason is that the laboratory measurement of these two properties (and hence the fixing of the referent of the variables) is always performed at equilibrium.

45 Ilya Prigogine. *From Being to Becoming* (New York: W. H. Freeman, 1980), pp. 6–8.

46 Ilya Prigogine and Isabelle Stengers. *Order Out of Chaos* (Toronto: Bantam, 1984), pp. 138–43.

47 Ilya Prigogine. *From Being to Becoming*. Op. cit., pp. 90–5.

The extremum states need not be steady states, but can be periodic or even chaotic. Like so many other advances in chemistry, the discovery of these other possibilities was stimulated by a new and problematic phenomenon: chemical reactions that displayed a *repetitive rhythm* in the concentration of the reacting substances. Given that in order for such a macroscopic time structure to emerge, millions of microscopic entities must somehow synchronize their activity, the existence of this periodic phenomenon was at first received with incredulity. But the phenomenon was not only remarkable but public and reproducible, forcing chemists to develop new mathematical tools to cope with it.

From personal to consensus practice

1 Bernadette Bensaude-Vincent and Isabelle Stengers. *A History of Chemistry* (Cambridge, MA: Harvard University Press, 1996), pp. 111–12.

2 Robert D. Purrington. *Physics in the Nineteenth Century*. New Brunswick, NJ: Rutgers University Press, 1997, p. 37.

Charles Coulomb had established in 1785 that electric forces had the mathematical form of an inverse-square relation. This important insight was extended by Siméon Denis Poisson who, using the mathematics of Pierre-Simon Laplace, created the first model for electrostatic phenomena in 1811–13.

3 Ian Hacking. *Representing and Intervening* (Cambridge, MA: Cambridge University Press, 1983), p. 243.

4 William H. Brock. *The Chemical Tree. A History of Chemistry* (New York: W.W. Norton, 2000), pp. 141–2.

5 Ibid., pp. 135–7.

6 Trevor H. Levere. *Affinity and Matter. Elements of Chemical Philosophy, 1800–1865* (Oxford: Clarendon, 1971), p. 36.

7 Ibid., pp. 37–9.

Of his contemporaries, only Jöns Jacob Berzelius could compete with Davy in the number of elementary substances isolated, and in the application of the new concepts, statements, and problems generated by electrolysis to models of affinity and chemical composition.

8 Ibid., p. 83.

9 Ibid., pp. 84–6.

Faraday had a close relation with William Whewell, a philosopher in collaboration with whom he developed the new terminology.

10 Ibid., pp. 87–8.

11 Ibid., pp. 95–8.

12 Hans-Werner Schütt. *Eilhard Mitscherlich: Prince of Prussian Chemistry* (Washington, DC: American Chemical Society, 1997), p. 34.

13 Ibid., p. 42.

14 Eric R. Scerri. *The Periodic Table. Its Story and Its Significance* (Oxford: Oxford University Press, 2007), pp. 42–3.

15 Ibid., p. 44.

Scerri argues that vertical regularities were easier to find at a time when atomic weights were calculated in the gaseous state. Horizontal regularities occur among elements separated by transitions from solid to gas.

16 Virginia M. Schelar. *Thermochemistry and the Third Law of Thermodynamics*. Chymia. Vol. 11 (Berkeley, CA: University of California Press, 1966), pp. 100–2.

17 Bernadette Bensaude-Vincent and Isabelle Stengers. *A History of Chemistry*. Op. cit., p. 121.

The statement implied by this empirical regularity, that all atoms have the same heat capacity, turned out to be only an approximation (it failed for diamond, and for all substances at very low temperatures), but within its limited range of validity it served as a useful means to bridge the macro and micro scales.

18 William H. Brock. *The Chemical Tree*. Op. cit., p. 359.

19 William Allen Miller. *Elements of Chemistry: Theoretical and Practical. Part 1. Chemical Physics* (London: John W. Parker, 1855).

20 John Murray. *A System of Chemistry*. Volume I (Edinburgh: Longman, Hurst, & Rees, 1806), p. 37.

21 Ibid., pp. 132–3.

22 William Allen Miller. *Elements of Chemistry*. Op. cit., pp. 8 and 56–7.

23 John Murray. *A System of Chemistry*. Op. cit., pp. 41–3.

24 William Allen Miller. *Elements of Chemistry*. Op. cit., pp. 93 (role of adhesion), 101–7 (crystal shape classification).

The triggering of crystallization in a saturated solution by the introduction of a seed crystal was already known as a phenomenon in 1800. But the explanation of it has been improved by the use of both adhesion and cohesion forces as part of the mechanism. On the phenomenon itself see: John Murray. *A System of Chemistry*. Op. cit., p. 44.

25 William Allen Miller. *Elements of Chemistry*. Op. cit., pp. 107–10.

In addition to isomorphism the author discusses two other concepts: *dimorphism* and *allotropy*. These refer to a phenomenon that is the very opposite of isomorphism: one and the same substance exhibiting various solid forms. Solid carbon, for example, can form octahedral crystals in diamond or rombohedral ones in graphite. Sulphur, too, can exhibit this dimorphism, as can a variety of compounds. The differences in solid form of one and the same substance can extend to alternative non- crystalline shapes, whether glass, clay, or fiber, alternative allotropes exhibiting different properties despite identical composition (ibid., pp. 111–14).

26 William Allen Miller. Op. cit., Part 2., pp. 1097–100.

This was Faradays' explanation.

27 Ibid., p. 1109.

28 Ibid., Part 1, pp. 324–5.

As mentioned in a footnote above, Faraday had a close relation with the philosopher William Whewell. They shared the conviction that a universal concept of polarity was the key to the unification of many scientific phenomena.

29 Ibid., p. 12 (statement 1), p. 69 (statement 2), pp. 351–2 (statement 3), p. 327 (statement 4), p. 335 (statement 5).

30 John Murray. *A System of Chemistry*. Op. cit.

Tables for these properties are given in the following pages: p. 62 (saturation), p. 174 (heat expansion), p. 178 (heat dilation), p. 266 (heat conductivity), p. 373 (heat capacity), p. 401 (latent heat), p. 490 (melting points).

31 William Allen Miller. *Elements of Chemistry*. Op. cit.

Tables for these properties are given in the following pages: Part 1, pp. 128–9 (refractive power), p. 275 (insulators and conductors), p. 298

(specific induction); p. 329 (electro-chemical order), p. 421 (magnetic power); and Part 2, p. 1123 (specific electricity).

32 Ibid., Part 1.

These tables appear in the following pages: p. 231 (latent heat), p. 168 (expansion of liquids by heat), p. 170 (expansion of gases by heat), p. 188 (heat conductivity).

33 Ibid., Part 2, p. 1124.

34 Ibid., pp. 1091–6.

35 Ibid., Part 1., p. 452 (atmosphere), p. 674 (glass), p. 759 (alloys); and Part 2., p. 1088 (mineral water).

36 Ibid., pp. 343–54.

The mathematical model in question is known as "Ohm's law."

37 Henry A. Laitinen and Galen W. Ewing. *A History of Analytical Chemistry* (Washington, DC: American Chemical Society, 1977), p. 104.

The phenomenon was discovered by William Hyde Wollaston, an English chemist and physicist, in 1802. The mapping of the details of the line's distribution in the solar spectrum was first done by Josef Fraunhofer in 1817.

38 William Allen Miller. *Elements of Chemistry*. Op. cit., p. 136.

39 Henry A. Laitinen and Galen W. Ewing. *A History of Analytical Chemistry*. Op. cit., pp. 106–8.

40 Eric R. Scerri. *The Periodic Table*. Op. cit., pp. 63–94.

The list of practitioners who worked on different versions of the Periodic Table (some using spirals or screws to display periodicity, others the more familiar arrangement of rows and columns) includes: Alexandre De Chancourtois, John Newlands, William Odling, Gustavus Hinrichs, Lothar Meyer, and Dimitri Mendelev.

41 Ibid., pp. 96, 109.

The recently discovered valency numbers, in turn, could be used as an aid in creating the groups along the second dimension. Both Meyer and Odling preceded Mendelev in the use of valency to correct the ordering by weight.

42 Ibid., pp. 113–18.

43 Ibid., p. 108.

Mayer anticipated Mendelev in predicting the existence of missing elements in his classification of 1862 (ibid., p. 96).

44 William H. Brock. *The Chemical Tree.* Op. cit., pp. 374–7.

A different set of quantitative dependencies was uncovered by François Raoult in work spanning several decades. Chemists already knew that when a substance is dissolved in a solvent the freezing point of the latter was depressed, the depression being proportional to the amount of solute. Raoult studied 60 different compounds dissolved in acetic acid, and showed that the depressions were constant, although organic and inorganic materials yielded different constants. This implied that the effect depended on the amount of solute and the identity of the solvent, but not on the chemical nature of the solute (ibid., pp. 362–5).

45 John W. Servos. *Physical Chemistry from Ostwald to Pauling. The Making of a Science in America* (Princeton, NJ: Princeton University Press, 1990), p. 28.

The concept of dynamic equilibrium was provided by Oliver Williamson; the means to measure equilibrium constants by Wilhelm Ostwald and others; and the equation by a pair of little known Danish chemists, Peter Waage and Cato Guldberg.

46 Alexander Kipnis. "Early Chemical Thermodynamics: Its Duality Embodied in Van't Hoff and Gibbs." In *Van't Hoff and the Emergence of Chemical Thermodynamics.* Edited by Willem J. Hornix and S. H. W. M. Mannaerts (Delft: Delft University Press, 2001), p. 6.

The author argues that van't Hoff exhibited great creativity in inventing ideal phenomena, that is, suitable cycles for the different applications of his model.

47 John W. Servos. *Physical Chemistry.* Op. cit., pp. 20–1 and 28–9.

48 Ibid., p. 37.

49 Ibid., pp. 3–4.

50 Ibid., p. 22.

The measurements of heats of reaction in neutralization reactions were performed by Julius Thomsen, the results published in 1869. Ostwald was aware of Thomsen's work, as well as that of others like Ludwig Wilhelmy who in 1850 had studied the relation between the velocity of a reaction and the quantity of substances reacting. Wilhelmy used a differential equation to model the progress of the reaction. See: Frederick L. Holmes. *From Elective Affinities to Chemical Equilibria: Berthollet's Law of Mass Action.* Chymia. Vol. 8 (Berkeley, CA: University of California Press, 1962), p. 130.

51 William H. Brock. *The Chemical Tree.* Op. cit., pp. 384–5.

Among the models he created, his "dilution law" stands out for its wider consequences. If we represent the portion of dissociated acid by a, the portion not dissociated by 1- a, and the volume of water in which the acid is dissolved by v, the model was:

$$a_2 / (1 - a) v = k$$

The constant k could be used to measure the relative strengths of acid and bases, and in the hands of biochemists, it became the basis for the pH scale in the early 1900s.

52 Ibid., p. 383.

Arrhenius, on the other hand, did not see a conflict between the two scales: his last contribution to chemistry was a study of reaction mechanisms through the concept of a molecule's activation energy.

53 Although for convenience we will use the expression "the reference textbook" we will use, in fact, three different textbooks, because we need a different reference textbook for physical and inorganic chemistry, and because the physical chemistry textbook is Ostwald's, his anti-atomism is on display throughout the book, so we need another text to compensate for this bias: the one written by Ostwald's first academic assistant, Walther Nernst. The three textbooks are: Wilhelm Ostwald. *The Fundamental Principles of Chemistry. An Introduction to All Textbooks of Chemistry* (New York: Longmans, Green, 1909); Walther Nernst. *Theoretical Chemistry. From the Standpoint of Avogadro's Rule and Thermodynamics* (London: MacMillan, 1904); and William Ramsay. *A System of Inorganic Chemistry* (London: J. & A. Churchill, 1891).

54 Wilhelm Ostwald. *The Fundamental Principles of Chemistry*. Op. cit., p. 6.

55 Walther Nernst. *Theoretical Chemistry*. Op. cit., pp. 286–8.

56 William Allen Miller. *Elements of Chemistry: Theoretical and Practical* (London: John W. Parker, 1855).

The word "caloric" does not appear in this text. And the text refers to new phenomena that suggest that the nature of heat is closer to that of light. In particular, by 1850 radiant heat has been found capable of being reflected (pp. 197–9), refracted, and even polarized (pp. 208–11).

57 Wilhelm Ostwald. *The Fundamental Principles of Chemistry*. Op. cit., pp. 315–33.

58 William Ramsay. *A System of Inorganic Chemistry*. Op. cit., p. 88.

59 John Murray. *A System of Chemistry*. Volumes I to IV (Edinburgh: Longman, Hurst & Rees, 1806), pp. 395–8.

60 Wilhelm Ostwald. *The Fundamental Principles of Chemistry*. Op. cit., pp. 17–18.

Ostwald uses the terms "quantities" and "intensities." The terms "extensive" and "intensive" are more recent names for these concepts. The definitions of these latter terms are identical today as the definitions of Ostwald's terms.

61 Walther Nernst. *Theoretical Chemistry*. Op. cit., p. 692.

Nernst acknowledges that this redefinition captures only the mode of action of affinity (its capacity to drive a flow of matter, for example), not its *nature*. The latter is presumably explained by directed affinities, as in stereochemistry (p. 278).

62 John Murray. *A System of Chemistry*. Op. cit., pp. 442 (mechanical equilibrium) and 342 (thermal equilibrium).

Even the electrical battery came from physics in 1800 with a concept of equilibrium attached to it: the balance of electrical forces in an electrolytic solution was said to be disrupted by contact with different metals (ibid., p. 592).

63 William Allen Miller. *Elements of Chemistry*. Op. cit., p. 191.

In 1800 thermal equilibrium was explained in terms of two mechanisms: conduction, the motion of heat through a material; and radiation, its motion through empty space (as in the transmission of radiant heat from the sun to the earth). In 1850, a new concept has been added, referring to a third mechanism for the transfer of heat and the establishment of equilibrium: *convection*, heat transferred by the motion of liquid or gaseous substances.

64 Walther Nernst. *Theoretical Chemistry*. Op. cit., pp. 429–30.

65 Wilhelm Ostwald. *The Fundamental Principles of Chemistry*. Op. cit., p. 84.

66 Ibid., p. 83.

67 Walther Nernst. *Theoretical Chemistry*. Op. cit., pp. 692–3 (Statements 3 and 4 appear on the same page).

68 Ibid., p. 27.

69 Ibid., p. 28.

70 Ibid., p. 17.

71 Wilhelm Ostwald. *The Fundamental Principles of Chemistry*. Op. cit., p. 33 (Statements 8 and 9 appear on the same page).

72 Ibid., p. 290.

73 Ibid., p. 293.

74 Ibid., p. 294.

75 Ibid., p. 295.

76 Ibid., p. 298.

77 William Allen Miller. *Elements of Chemistry*. Op. cit., pp. 435–6.

78 William Ramsay. *A System of Inorganic Chemistry*. Op. cit., pp. 22–3.

When this textbook was published, the Periodic Table was still under construction. It is missing an entire column (the inert gases) and the rare earths have not found their place yet. But there is enough regularity in the order of the lighter elements for the taxonomic scheme to seem correct, and more importantly, to seem improvable.

79 William H. Brock. *The Chemical Tree*. Op. cit., p. 386.

80 Walther Nernst. *Theoretical Chemistry*. Op. cit., p. 498.

These problems do not appear as "Why" questions, but as statements preceded by phrases like "It is not known why ..." or followed by "the causes of which are unknown."

81 Ibid., pp. 499–500.

In these pages Nerst does not examine the question why but *how* the degree of dissociation depends on chemical nature, and enumerates a few empirical rules. But the causes remain unknown, as the author notes, particularly in the case of electrolytes that cleave into more than two ions.

82 Wilhelm Ostwald. *The Fundamental Principles of Chemistry*. Op. cit., p. 297.

Ostwald acknowledges the effect of temperature of reaction rates but also the fact that all the factors determining velocity are to a large extent unknown.

83 Ibid., p. 299.

84 Ibid., p. 298.

85 Ibid., p. 293.

86 This is certainly true for Ostwald's textbook, in which the diagrams are used throughout the book. But they are also prominent in Nernst's textbook which provides two detailed analyses of chemical reactions using phase diagrams: Walther Nernst. *Theoretical Chemistry*. Op. cit., pp. 609–15, 625–8.

87 Wilhelm Ostwald. *The Fundamental Principles of Chemistry*. Op. cit., p. 305.

88 Walther Nernst. *Theoretical Chemistry*. Op. cit., p. 614.

89 Wilhelm Ostwald. *The Fundamental Principles of Chemistry*. Op. cit., p. 234.

The status of phase diagrams as cognitive tools is not clear. On the one hand, they may be considered descriptive devices, maps of exhaustively enumerated combinations of thermodynamic properties. On the other hand, the *structure* of the space of possibilities may be considered to play an explanatory role. The kind of explanation the diagrams supply, however, is not causal, since they do not specify any mechanism. The term "quasi-causal" may be used to characterize their explanatory role. A similar point was made in the previous section about the other forms of stability (periodic, chaotic) that were added to a dynamic steady-state once the thermodynamic treatment was extended far from equilibrium. A detailed treatment of the causal and quasi-causal components of an explanation, from physics to economics and linguistics, can be found in: Manuel DeLanda. *Philosophy and Simulation. The Emergence of Synthetic Reason* (London: Continuum, 2011).

Chapter 4: Social Chemistry

1 Stevin Shapin and Simon Schaffer. *Leviathan and the Air Pump. Hobbes, Boyle, and the Experimental Life* (Princeton, NJ: Princeton University Press, 1987), pp. 4–7.

In this influential study, for example, laboratory culture is assumed to have existed since the seventeenth century, and to have maintained its coherence as a world-view for 400 years, to the extent that any twentieth-century historian who makes an effort to understand the cognitive components of a field is assumed to be inextricably linked to this culture. Given this presupposition, the only methodologically viable strategy is to find a way to play an outsider, by locating a historical controversy in which a real outsider participated, and then taking the outsider's point of view. Another way of following this strategy is by studying a contemporary laboratory as if its inhabitants were members of a primitive tribe, while studiously avoiding the temptation to become one of the natives, as is done in: Bruno Latour and Steve Woolgar. *Laboratory Life. The Construction of Scientific Facts* (Princeton, NJ: Princeton University Press, 1987), pp. 37–8.

In both cases what seems indispensable to the methodology is a deliberate avoidance of any information concerning the cognitive content of a field, since any exposure to that information would transform the researchers into insiders.

2 Larry Laudan. *Beyond Positivism and Relativism. Theory, Method, and Evidence* (Boulder, CO: Westview Press, 1996), pp. 7–22.

Laudan argues that empiricists (positivists) and constructivists (relativists) share many assumptions. In addition to their common holism and conventionalism, both camps believe that the content of science is basically linguistic, so that evidence can be reduced to observation statements, and conflict between rival theories reduced to their statements being untranslatable. Even where they disagree, they do so over a similar set of alternatives: there must be either infallible rationality or there is no rationality at all; and either all cognitive content is retained after a controversy or no content is retained at all.

3 The term "theory" has been avoided in this book because it implies that the set of cognitive tools shaping personal and consensus practices possesses more homogeneity than it actually does. And similarly for the term "observation," which does not do justice to the variety of practices that go on in real laboratories. We will make an exception in this chapter because the debate over underdetermination is cast in these terms: the relation between theoretical and observational statements. Nevertheless, and strictly speaking, every occurrence of these terms in the paragraphs that follow should be inside quotation marks, an expedient avoided only for aesthetic reasons.

4 Brian Ellis. "What Science Aims to Do." In *Images of Science*. Edited by Paul M. Churchland and Clifford A. Hooker (Chicago: University of Chicago Press, 1985), p. 64.

In this model a theory is a set of statements of the form:

$$C_1 \text{ and } C_2 \text{ and } .... \text{ Cn entail O}$$

in which the Cs are empirically determined boundary conditions and O is an observation statement. Ellis argues that the model is so widespread that it might very well be called "the standard model." Others refer to it as "the received view."

5 Willard Van Orman Quine. "Two Dogmas of Empiricism." In *From a Logical Point of View* (New York: Harper and Row, 1963), pp. 42–4.

This line of argument is known as the Duhem-Quine thesis. Pierre Duhem is the chemist and positivist philosopher who first suggested the argument.

6 Larry Laudan. *Beyond Positivism and Relativism*. Op. cit., pp. 50–2.

7 Barry Barnes, David Bloor, and John Henry. *Scientific Knowledge. A Sociological Analysis* (Chicago: University of Chicago Press, 1996), pp. 71–2.

This approach is called *finitism*. It is based on five principles, two of which state that the past meanings of words underdetermine the meanings they can have in the future, and vice versa, that all past uses are revisable (Principles 1 and 3). From these two it follows that no set of statements (beliefs, classifications) is indefeasibly true or false (Principle 2). We can accept these three principles but with two modifications. First, we must not assume that meaning determines reference, as the authors do. And second, the principles must be used carefully to avoid self-referential incoherence: if one asserts that given the fluidity of meanings one cannot hope to specify *even at a given instant* what a theory implies (p. 93) then the very principles of finitism (and their implications) are called into question. Principles 4 and 5 are more objectionable: they assert the *holistic* character of the set of statements such that modifying one member of the set has repercussions for all the others. This, of course, implies the acceptance of relations of interiority and the denial of the partial autonomy that the components of a "theory" have. It must be added that, strictly speaking the authors do not adopt a conventionalist solution to the underdetermination problem. The reason is that, if conventions are conceived as explicit rules, then the stability of their meaning (and hence their capacity to define activities) is subject to finitist doubts (p. 55). Thus, they replace logical compulsion not by a conventional choice among alternatives, but by *sociological compulsion*: the professional interests and goals furthered by picking one alternative over another (p. 120). This conclusion, of course, also runs the risk of self-referential incoherence: it may have been reached only because it furthers the professional interests of sociologists.

8 Philip Kitcher. *The Advancement of Science. Science without Legend, Objectivity without Illusions* (New York: Oxford University Press, 1993), pp. 209 and 249–52.

9 Chemists in 1700 using the term "vitriol" and those in 1800 using "sulphuric acid" had different conventional standards of purity, so the samples of substances they referred to were probably not exactly the same. Nevertheless, as long as the chemical reactions into which the substances of different purity entered were not affected by this, the differences could be considered insignificant. It could be objected that the very fact that standards of purity are needed to decide when two samples of a substance are the same proves that conventions are involved. This is correct, but harmless if the conventions in question are coordinative and not constitutive, as argued below in the main text. On the question of fixing the referent of the term "heat" by its effects see: John Murray. *A System of Chemistry*. Vol. I (Edinburgh: Longman, Hurst & Rees, 1806), p. 144.

10 Philip Kitcher. *The Advancement of Science*. Op. cit., p. 247.

11 Brian Ellis. "What Science Aims to Do." Op. cit., pp. 65–6.

12 Andrei Marmor. *Social Conventions. From Language to Law* (Princeton, NJ: Princeton University Press, 2009), p. 10.

13 Ibid., p. 8.

14 Edwin C. Kemble. *Physical Science. Its Structure and Development* (Cambridge, MA: MIT Press, 1966), p. 423.

15 Andrei Marmor. *Social Conventions*. Op. cit., pp. 31–6.

16 It could be argued that the convention "Celsius units" constitutes the identity of the practice "to measure temperature with Celsius units." But this would be trivial in that the practice would not be significantly different from that of measuring temperature with other units. And moreover, the identity of the referent of the term "temperature" is not constituted by the choice of units. There are mathematical techniques (dimensional analysis) that can take equations in which the variables are expressed in conventional units and convert them into equivalent equations using no units at all. See: Jaap van Brakel. *Philosophy of Chemistry* (Leuven: Leuven University Press, 2000), pp. 175–77.

17 Andrei Marmor. *Social Conventions*. Op. cit., p. 45.

18 Ibid., p. 37.

19 Stevin Shapin and Simon Schaffer. *Leviathan and the Air Pump*. Op. cit.

The authors argue that "a fact is a constitutively social category" and that conventions are used to define the means to generate facts (p. 225). They use the term "fact" to refer to what in this book has been called an "objective phenomenon." When we introduced this term we said that phenomena had to be publicly recognizable, recurrent, and noteworthy. The constitutive conventions in question are those that determine what counts as being public and recurrent, that is, replicable by others. In the case analyzed by the authors (the use of air pumps to study pressure phenomena, respiration phenomena, and so on) the public status of a phenomenon was ensured by a ritualistic use of trustworthy witnesses (pp. 56–8) and by the propagation of drawings of phenomena, like an illustration of a dead mouse in a vessel from which air had been evacuated (pp. 60–2). Now, we can accept that *in this case* the means to establish trustworthiness were defined by a convention: what constituted a reliable witness was both knowledgeability and the presumed moral fiber of the upper class. The problem is that neither the ritual itself nor the use of drawings to create "virtual witnesses" was a constituting feature of all laboratories. In chemical laboratories, arguably a more important case than that of physics laboratories until the nineteenth century, there were neither spectacular displays to a group of gentlemen (as opposed to experts in the field), nor any significant role for virtual witnessing through published illustrations.

This leaves only the case of the replicability of a phenomenon. The authors study two decades, 1650–70, in which the population of air pumps had only a few members, so few indeed that technical problems (such as leaks) could not be reliably ascribed to any one factor. In those conditions, there was circularity in the definition of what constituted a phenomenon: the latter was defined as real if it was replicable using the air pump, while a correctly functioning air pump was defined as one in which the phenomenon could be replicated. The authors acknowledge, however, that in the following decade commercial pumps appeared, greatly increasing the size of the population of instruments and making many problems unambiguously diagnosable (pp. 274–6). We may conclude that conventions played a role in constituting what counted as replicable only when air pumps were rare and imperfect, but it can be argued that during that brief period a phenomenon *has not yet become part of the domain*, since it has not become public.

20 Harry Collins. *Changing Order* (Chicago: University of Chicago Press, 1985).

Collins refers to situations in which there is vicious circularity as involving an *experimental regress*. He shows that such a regress exists in a single case: gravity wave detectors (pp. 83–4). But in other cases, such as that of a TEA laser, the phenomenon itself is so remarkable and easy to witness, and its capacities (such as the ability to cut through solid materials) so easy to establish, that no experimental regress exists (p. 74). The only generalizable point about his other cases is that the replication of a phenomenon is not something that can be performed by using an infallible mechanical recipe (an algorithm). This is true but trivial, that is, true but only when compared with the picture of what is involved in testing a theory given by the standard model.

21 David Lewis. *Convention* (Oxford: Blackwell, 2002), p. 8.

22 Not everyone's preferences were like this. Berthellot went on using equivalents even after the1860 agreement because his distaste for atomic weights was stronger than his desire for coordination. His authority in Paris was so great that he became an obstacle for the propagation of the agreed standards in France. See: Mary Jo Nye. *From Chemical Philosophy to Theoretical Chemistry. Dynamics of Matter and Dynamics of Disciplines* (Berkeley, CA: University of California Press, 1993), p. 143.

23 David Lewis. *Convention*. Op. cit., p. 25.

24 Ibid., pp. 84–6.

Lewis emphasizes this point, that an explicit agreement is not necessary to solve a coordination problem, because he intends his model to serve in the case of the emergence of language: assuming the need for agreement in this case makes the argument circular, because language is needed to make agreements explicit.

25 Bernadette Bensaude-Vincent and Isabelle Stengers. *A History of Chemistry* (Cambridge, MA: Harvard University Press, 1996), pp. 136–9.

26 Brian Ellis. "What Science Aims to Do." Op. cit.

Ellis makes a distinction between causal theories and abstract model theories. Concepts appearing in statements of the former refer to real entities with causal powers that can be exercised in a number of ways, and therefore entities that can be checked for their existence in different ways (p. 64). The entities appearing in statements of the latter, on the other hand, are not postulated as causes and are not assumed to have any effects (p. 58). Ideal gases and ideal engines are examples of these type of entities. He concludes that conventions may be constitutive of abstract model theories (such as theories of space-time) but not of causal theories (p. 66). His distinction is similar to the one drawn here: as argued in Chapter 3, equations describing significant correlations between the variables defining an ideal phenomenon *pose a problem*, a problem that must be subsequently explained using a causal explanatory schema or model.

27 Andrei Marmor. *Social Conventions*. Op. cit., pp. 14–15.

28 A full discussion of the distinction between properties and capacities in different scientific fields is given in: Manuel DeLanda. *Philosophy and Simulation. The Emergence of Synthetic Reason* (London: Continuum, 2011).

29 Adolphe C. Wurtz. *An Introduction to Chemical Philosophy According to the Modern Theories* (London: J.H. Dutton, 1867), pp. 92–3.

Wurtz was an atomist, so he explains capacities to react in molecular terms: type formulas are models that show the different *directions along which molecules are capable of breaking* (p. 91).

30 When placing a scientific field in its social context the main danger to be avoided is to think that there are only two relevant social scales: the micro-level of individual practitioners and the macro-level of society as a whole. Sociologists who subscribe to methodological holism tend to think of this macro-level in terms of relations of interiority, and view everything below it as determined by its function in the overall totality. Using relations of exteriority, on the other hand, multiple intermediate scales can be defined comprising social entities like communities, institutional organizations, and cities. These can be conceived as wholes with properties of their own, thereby blocking a reductionist treatment while at the same time preserving the relative autonomy of each entity. An account of social ontology that avoids the methodological holism of macro-sociologists without yielding to the methodological individualism of micro-economists is given in: Manuel DeLanda. *A New Philosophy of Society* (London: Continuum, 2006).

31 Stephen Toulmin. *Human Understanding. The Collective Use and Evolution of Concepts* (Princeton, NJ: Princeton University Press, 1972), p. 146.

32 Mary Jo Nye. *From Chemical Philosophy to Theoretical Chemistry. Dynamics of Matter and Dynamics of Disciplines, 1800–1950* (Berkeley, CA: University of California Press, 1993), p. 29.

33 Ibid., p. 29.

The author defines the institutional identity of a discipline by six elements: a real genealogy and myths of heroic origins; a core literature containing a shared language; practices and rituals; a physical homeland; external recognition; and shared cognitive values and problems. But she emphasizes that the last element, the cognitive content, far from being the end result of genealogy, ritual, and ceremony, is the starting point for them.

34 Bernadette Bensaude-Vincent and Isabelle Stengers. *A History of Chemistry*. Op. cit., pp. 63–4.

35 Mi Gyung Kim. *Affinity, That Elusive Dream. A Genealogy of the Chemical Revolution* (Cambridge, MA: MIT Press, 2003), pp. 17–18.

36 Bernadette Bensaude-Vincent and Isabelle Stengers. *A History of Chemistry*. Op. cit., p. 65.

37 Mi Gyung Kim. *Affinity*. Op. cit., p. 221.

Like other French chemists, Macquer praised the writings of Stahl, filled with useful chemical operations and procedures, arguing that they gave a truer picture of the nature of chemistry than the abstract concepts of the natural philosophers. Promoters of chemistry in England also faced the choice of emphasizing the prestigious pure science aspects, or the useful applied aspects, well into the nineteenth century, different chemists finding different ways to solve the tension between the two. See: Robert Friedel. "Defining Chemistry: Origins of The Heroic Chemist." In *Chemical Sciences in the Modern World*. Edited by Seymour H. Mauskopf (Philadelphia, PA: University of Pennsylvania Press, 1993), pp. 219–21.

38 Barry Gower. *Scientific Method. An Historical and Philosophical Introduction* (London: Routledge, 1997).

Gower argues that since Galileo, the uncertainties associated with evidence from experiments made the use of rhetoric indispensable in the presentation and justification of general statements (p. 23). This may sometimes work against the development of good science, but other times, when the evidence is elusive and difficult to identify, it may play a constructive role, guiding researchers to the most reasonable (but not compelling) conclusion (p. 241).

39 Larry Laudan. *Beyond Positivism and Relativism*. Op. cit., pp. 211–12.

40 Barry Gower. *Scientific Method*. Op. cit.

These two basic positions towards certainty (deductive and inductive) were adopted by Galileo (p. 24) and Bacon (p. 53) respectively. Descartes accepted eliminative induction but conceded that the certainty it delivered was moral or practical, not apodictic (p. 66).

41 Ibid., p. 46.

Unlike Galileo, Bacon wanted to break with Aristotle, and in particular with his distinction between science and craft. Far from being foreign to science, the activities of cooks, weavers, farmers, carpenters, apothecaries had to be investigated and "natural histories" written to catalog the effects they were able to produce.

42 Ibid., pp. 80–1.

Gower argues that Newton's famous opposition to the postulate hypothetical causes stemmed from his attempt to reconcile the conflict between, on the one hand, retaining a sharp divide between knowledge and opinion, and on the other hand, giving up apodictic certainty and accepting only the practical certainty that laboratory evidence could give. Because practical certainty comes in degrees, it makes the border with opinion fuzzy, so Newton used his rigid position on hypotheses as a rhetorical means to ameliorate the impact of the blurred boundary.

43 Barry Barnes, David Bloor, and John Henry. *Scientific Knowledge*. Op. cit., Chapter 6.

In their chapter on drawing boundaries, the authors uncritically agree with Stevin Shapin and Simon Schaffer that the criterion of what counts as science was given by a few methodological remarks made by Robert Boyle in the seventeenth century, remarks like "causes can be debated but not facts," or something is scientific if it is "consonant with observation and experiment." These are taken to be constitutive conventions. But as both the authors and the historians they admire admit, Boyle violated his own rules on several occasions, much as Newton violated his own strictures against not framing hypotheses. One may wonder what kind of constitutive convention can those rhetorical remarks be if they are not binding on the very authors who proclaimed them.

44 Joseph Ben-David. *The Scientist's Role in Society* (Chicago: University of Chicago Press, 1984), p. 78.

Ben-David explores the question of why some cities or countries become dominant scientific centers, replacing previous centers and, in time, being replaced by others. He explains the particular case of the change in dominance between England and France in the eighteenth century by the role played by their respective scientistic movements (ibid., p. 83).

45 Tom Sorell. *Scientism. Philosophy and the Infatuation With Science* (London: Routledge, 1994), pp. 4–5.

46 Richard G. Olson. *Science and Scientism in Nineteenth-Century Europe* (Chicago: University of Illinois Press, 2008).

Henri Saint-Simon, founder of an important scientistic movement, seems to have believed in a Newtonian social science, in which Newton's third law could be rendered as "One can only be truly happy in seeking happiness in the happiness of others" (ibid., p. 46). The leaders of other scientistic movements, such as Pierre-Jean George Cabanis and Auguste Compte, believed that each field had its own domain of application and was not reducible to physics, but were nevertheless unifiable (ibid., pp. 23, 63). Some thought that the fact that the scientific method endowed results with an objective status was enough to compel their acceptance, while others realized that these results had to be combined with a natural religion to be truly persuasive: Saint-Simon offered New Christianity as a scientifically-based alternative to Catholicism, while Compte dreamed of a Religion of Humanity and a Church of Positivism (pp. 52, 65).

47 Ibid., pp. 58, 67, 153.

48 Joseph Ben-David. *The Scientist's Role in Society.* Op. cit.

This is, of course, a famous remark by Galileo. We do not mean to suggest that this remark achieved much by itself, because the dynamics through which scientific fields acquired legitimacy was complex and varied, and it involved interactions between a variety of communities and organizations. Relations with religious organizations, for example, varied depending on whether or not they had a centralized authority, and whether their doctrine was rigid or fluid: Protestants were more receptive to the values of natural philosophy than Catholics, but only when they did not exist in small isolated communities (pp. 71–2). The relations with governmental organizations also varied according to their stability and their attitude towards debate and dissent: in countries like mid-seventeenth-century England, where debates about religion and politics had reached an ideological impasse, the Baconian values of natural philosophy spread more widely because it was believed they provided a neutral meeting ground (p. 74).

49 Barry Gower. *Scientific Method.* Op. cit., p. 136.

The philosopher in question was Emmanuel Kant, who famously claimed that it was possible to have a priori knowledge of matters of fact, that is, knowledge that was synthetic a priori. He used classical mechanics, Euclidean geometry, and the Aristotelian syllogism as his examples. A century later, all three turned out not to be a priori.

50 Ibid., p. 72.

The expression "deduction from the phenomena" was used by Newton to explain how he had derived his inverse square law.

51 Jaap van Brakel. *Philosophy of Chemistry.* Op. cit., pp. 151–2.

52 Nancy Cartwright. *How the Laws of Physics Lie* (Oxford: Clarendon Press, 1983).

Cartwright distinguishes between phenomenological (or causal) laws that explain how a phenomenon is produced, and fundamental (or abstract) laws that fit a phenomenon into a general framework, and whose role is to unify rather than to explain (pp. 10–11). Fundamental laws do not govern objects in reality, only objects in models (p. 129) and involve conventions in the form of a *ceteris paribus* clause that specifies the simplified conditions under which the law is true (p. 58).

53 Ibid., p. 11.

While we accept Cartwright's conclusions, it can be argued that the unifying power of mathematical laws is itself in need of explanation, and that when this account is properly given they do turn out to possess ontological content. In particular, the *topological invariants* of the models expressing those generalizations (their dimensionality, their distribution of singularities) do refer to something real: the structure of possibility spaces defining objective tendencies and capacities. For a full discussion of this point see: Manuel DeLanda. *Intensive Science and Virtual Philosophy* (London: Continuum, 2001), Chapter 4.

54 This reduction of data about real properties to the observation of the markings in a measuring instrument is also performed by constructivists. Latour and Woolgar, for example, view the activity of measurement exclusively in terms of the numerical (or graphic) output of different instruments. They go further than positivists in this reduction, however, because they disregard the semantic content of the output (numerical values, significant features of a graph), reducing it to mere *inscriptions*. See: Bruno Latour and Steve Woolgar. *Laboratory Life*. Op. cit., pp. 49–52.

55 Kathryn M. Olesko. "The Meaning of Precision: The Exact Sensibility in Early Nineteenth-Century Germany." In *The Values of Precision*. Edited by M. Norton Wise (Princeton, NJ: Princeton University Press, 1995), p. 109.

The invention of the Least Squares method is credited to the German mathematician and physicist Johann Carl Frederick Gauss, who used it in 1801 to model astronomical observations. He published it in 1809. He shares the credit with the French mathematician Adrien-Marie Legendre, who published a similar method in 1805.

56 Zeno G. Swijtink. "The Objectification of Observation: Measurement and Statistical Methods in the Nineteenth-Century." In *The Probabilistic Revolution*. Volume 1. Edited by Lorenz Krüger, Lorraine J. Daston, and Michael Heidelberger (Cambridge, MA: MIT Press, 1990), p. 265.

57 Ibid., p. 272.

58 Frank Wigglesworth Clarke. "The Constants of Nature: A Recalculation of the Atomic Weights." *Smithsonian Miscellaneous Collections*. Volume 54, Number 3 (Washington, DC: Smithsonian Institution, 1910), p. 3.

The author began his collection of all data about atomic weights in 1877, and completed the statistical modeling in 1882.

59 Jan Golinski. "The Nicety of Experiment: Precision of Measurement and Precision of Reasoning in Late Eighteenth-Century Chemistry." In *The Values of Precision*. Op. cit., pp. 76–86.

60 Zeno G. Swijtink. "The Objectification of Observation." Op. cit., p. 273.

61 John R. Taylor. *An Introduction to Error Analysis* (Sausalito, CA: University Science Books, 1997), pp. 13–16.

62 Kathryn M. Olesko. "The Meaning of Precision." Op. cit., pp. 113–15.

63 M. Norton Wise. "Precision: Agent of Unity and Product of Agreement." In *The Values of Precision*. Op. cit., pp. 227–8.

Wise notes that the implementation of this task varied from country to country: England, where trustworthiness was identified with class-extraction, was less receptive to the use of Least Squares than Germany, where trust was linked not to the gentlemanly status of witnesses but to the methodical exposure of errors to public scrutiny. These moral values were needed not only to justify the elimination of constant errors, but more generally, to legitimize uniform standards of measurement, a task to which the governments of different countries had applied themselves since the eighteenth century. Finally, another source of rhetoric was the use of this methodology in the social sciences. For example, Adolphe Quetelet dreamed of a social physics made possible by the calculus of probabilities. He compared the results of anatomical measurements of soldiers to those coming out of laboratories or observatories, arguing that in both cases the method of Least Squares gave the most probable value. But the argument he gave for the analogy was mostly rhetorical: he claimed that the distribution of errors in observations was, like the distribution of variation in anatomical traits, produced by a multitude of small deterministic events, a claim that was later shown to be wrong. See: Ian Hacking. *The Taming of Chance* (Cambridge: Cambridge University Press, 2012), pp. 108–13.

64 Tom Sorell. *Scientism*. Op. cit., pp. 5–8.

Sorell argues that the five analogs to the basic scientistic statements characterize the philosophical movement known as the Vienna Circle (or logical positivism). He speaks not of the "standard model" but of the "received view" on theories.

65 Bas Van Fraassen. *Laws and Symmetry* (Oxford: Clarendon Press, 1989), p. 222.

Van Fraasen is the leader of the so-called *semantic approach* to theories, an approach in which language becomes largely irrelevant to the subject. He urges philosophers to return to the mathematical tools (differential equations) actually used by practicing scientists. There are, on the other hand, several ways of carrying out this program. For a critique of Van Fraasen's approach and a description of an alternative way of defining the cognitive content of mathematical models, in terms of singularity distributions and invariants under transformations, see: Manuel DeLanda. "Deleuze, Mathematics, and Realist Ontology." In *The Cambridge Companion to Deleuze*. Edited by Daniel W. Smith and Henry Somers-Hall (Cambridge: Cambridge University Press, 2012), pp. 220–36.

66 Deborah G. Mayo. *Error and the Growth of Experimental Knowledge* (Chicago: University of Chicago Press, 1984), p. 129.

Mayo shows that the relation between theory and observation involves not a deductive relation between sets of statements but a series of models connected non-deductively: a model of a primary scientific hypothesis; a model of the data; and a model of the experiment itself to link the former two models together. Historical evidence that this is indeed the way theory and data were connected in Germany starting in the middle of the nineteenth century can be found in: Kathryn M. Olesko. *Physics As a Calling. Discipline and Practice in the Königsberg Seminar for Physics* (Ithaca, NY: Cornell University Press, 1991).

67 Stephen Toulmin. *Human Understanding. The Collective Use and Evolution of Concepts* (Princeton, NJ: Princeton University Press, 1972), pp. 206–9.

68 Ibid., pp. 225–6.

69 Larry Laudan. *Science and Values. The Aims of Science and Their Role in Scientific Debate* (Berkeley, CA: University of California Press, 1984), pp. 62–3.

Laudan calls this model, with independent levels and a two-way flow of justification, *the reticulated model*. The three levels he calls the factual, methodological, and axiological levels of justification. We adhere to his presentation here with two exceptions: he applies the model to entire theories, while in this book the notion of "the truth of a theory" is not accepted as valid; and he applies it to science as a whole, instead of to relatively autonomous fields, like chemistry. This leads him to pick examples that are too general (to apply across fields) and to disregard the fact that criteria for rightness of fit must be more local. Thus, he illustrates the axiological level with goals like "striving to save the phenomena" as opposed to "aiming at discovering the causes behind phenomena" (p. 48), while the level of methodology is exemplified by rules like "choose the rival theory with novel predictions" as opposed to "pick the one with most supporting evidence, predictive or retrodictive" (p. 34).

70 Ibid.

The reticulated model, in which the factual, methodological, and axiological levels co-evolve, can be used to explain *convergent belief change*, that is, consensus formation (p. 23), but it must also be able to explain episodes of dissensus in the historical record: the rules must locally underdetermine the choice of facts (pp. 26-7) and the goals must locally underdetermine the choice of rules (pp. 35-6).

71 Ursula Klein. *Experiments, Models, and Paper Tools* (Stanford, CA: Stanford University Press, 2003), p. 268.

72 Ibid., p. 204.

73 Ibid., p. 205.

Klein gives as examples Dumas and Liebig, who thought that the possibility of adjusting the different items to one another increased the probability that the approach was correct.

74 Ian Hacking. "The Self-Vindication of Laboratory Sciences." In *Science As Practice and Culture*. Edited by Andrew Pickering (Chicago: University of Chicago Press, 1992), pp. 54-5.

This particular rule of evidence is discussed by a variety of authors in this volume. In addition to Hacking, there is Andrew Pickering's dialectic of resistance and accommodation, as well as David Gooding's asymptotic realism. The only difference between their approaches and the one taken here is that they postulate this convergence by mutual adjustment as a general mechanism for all laboratory science, whereas here it is used to explain only particular episodes in the history of individual fields.

75 Larry Laudan. *Science and Values*. Op. cit., pp. 55-9.

Laudan argues that the slow acceptance of the hypothetico-deductive method illustrates a situation in which explicit goals (to save the phenomena) entered into conflict with implicit goals, such as discovering causal mechanisms behind the production of phenomena. He uses examples from natural philosophy. But chemistry seems to present a more complex case, one in which description, causal explanation, and classification coexisted as explicit goals. Chemists who, as part of explicit programmatic statements, were in favor of a search for causes would simply abandon that goal when it proved too difficult to achieve. See: Mary Jo Nye. *From Chemical Philosophy to Theoretical Chemistry*. Op. cit., pp. 59–68.

76 Alan J. Rocke. *Image and Reality: Kekule, Kopp, and the Scientific Imagination* (Chicago: University of Chicago Press, 2010), pp. 12–21.

Rocke analyzes the acceptance of the hypothetico-deductive method in the case of type formulas, as interpreted structurally by Williamson. In this case, the successful prediction was not about isomers but about models of the chemical reaction that produced ether from alcohol and sulphuric acid. On the other hand, Gerhardt, whose type formulas were used by Williamson and who was quite pleased with the results, did not consider the evidence sufficient to compel assent for the actual existence of atoms arranged in space (p. 33).

77 Peter J. Ramberg. *Chemical Structure, Spatial Arrangement: The Early History of Stereochemistry 1874–1914* (Aldershot: Ashgate, 2003), p. 243.

Ramberg stresses the fact that for anti-atomist chemists like Fischer, who investigated glucose and synthesized many of its isomers, structural formulas were simply an isomer-counting device, useful to classify sugars but carrying no ontological commitment to atoms.

78 Riki G. A. Dolby. *Debates Over the Theory of Solution: A Study of Dissent in Physical Chemistry in the English-Speaking World in the Late Nineteenth and Early Twentieth Centuries*. Historical Studies in the Physical Sciences. Vol. 7 (Berkeley, CA: University of California Press, 1976), pp. 310–11.

79 Ibid., pp. 327–8.

The rival model of liquid solutions was based on the idea that even if salt molecules become dissociated they should immediately interact with water molecules forming compounds (hydrates), a hypothesis that seemed to be backed by measurements of the changing properties of water at different degrees of concentration of the solute performed by Mendelev.

80 Stephen Toulmin. *Human Understanding*. Op. cit., pp. 263–4.

81 Joseph C. Hermanowicz. *Lives in Science: How Institutions Affect Academic Careers* (Chicago: University of Chicago Press, 2009), p. 233.

82 But if the reference group of a nineteenth-century chemist contained Boyle, Lavoisier, and Dalton—none of whose personal practices were representative in their time—and if the chemist thought of this group in mythologized terms, then this reference group would not contribute to the diffusion of novel or better cognitive tools. On the question of how representative the practices of different "founding fathers" are, see: Mi Gyung Kim. *Affinity, That Elusive Dream. A Genealogy of the Chemical Revolution* (Cambridge, MA: MIT Press, 2003), pp. 439–45.

83 William H. Brock. *The Chemical Tree. A History of Chemistry* (New York: W. W. Norton, 2000), p. 378.

84 Stephen Toulmin. *Human Understanding*. Op. cit., p. 280.

85 Ibid., pp. 268–9.

Toulmin does not use the expression "relations of exteriority," but the concept is implied in the argument that sociologists make a mistake when dealing with the social infrastructure of scientific fields as a single coherent system. Toulmin grants that this may be true of the roles *within* an institution, which are often systematically related and defined by their mutual relations, but he argues that it is an error to extrapolate from this case to the institutional population, in which links between organizations are less fixed and can change independently.

86 Ibid., pp. 270–1.

87 Ibid., pp. 279, 282–93.

Toulmin estimates that tracking changes in reference groups could be done by sampling the population of organizations and the community of practitioners every five years. Clearly, more empirical data is needed to establish relevant time-scales for change, and to decide whether these time-scales have remained stable over time or whether change has been accelerated in recent decades.

88 Ibid., p. 278.

89 Riki G. A. Dolby. *Debates Over the Theory of Solution*. Op. cit., pp. 343–5, 386–7.

The participants in the debate were, on the side of physical chemistry, van't Hoff, Arrhenius, and Ostwald. On the other side there were Henry E. Amstrong, P. Spencer Pickering, and Louis Kahlenberg. The debate began in 1887, when physical chemists began publishing their own journal and brought international attention to the main ideas. It raged in Britain through the 1890s, and was continued in the United States by Kahlenberg in the first decade of the twentieth century. By 1907, at the meeting organized by the Faraday Society, the debate was still going on but it had changed nature, with new intermediate positions developed.

90 Ibid., pp. 298, 332–7, 360.

91 Ibid., pp. 322, 349.

Two scientists played this moderating role, the chemist William Ramsay and the physicist Oliver J. Lodge.

92 Ibid., p. 392.

Dolby argues that there was no clear winner. Physical chemistry prevailed, but its main defender, Ostwald, was an anti-atomist, that is, the wrong kind of spokesman to promote the acceptance of the existence of charged ions. The latter were accepted not because of anything that he argued, but on the basis of the new ideas about atoms developed by chemists and physicists in the early twentieth century. Moreover, the mathematical models that prevailed were not those that the physical chemists had themselves developed, but the more elegant and general ones created by mathematical physicists like Gibbs.

93 Philip Kitcher. *The Advancement of Science*. Op. cit., p. 288.

INDEX OF AUTHORS

INDEX OF SUBJECTS

complete/incomplete 97–8, 105, 107
displacement 3, 18, 25–6, 30, 33, 45, 56, 97
electrochemical 99, 108–9, 114–16, 120–1, 126–7, 132
energy changes in 104–5
models of 59, 74, 75–6, 78, 84
neutralization 46, 99, 116, 118, 127
as phenomena 39, 43, 97, 104, 114, 125
substitution 56–7, 76–7, 79, 83, 89, 138
velocity of 131–2
relations of exteriority/interiority 135–8, 144, 151, 156
and holism ix, 11, 135–6, 151
rhetoric (myth, propaganda) xi, 144–6, 149
and scientistic discourse 146–7, 150

scientific field ix, 144–5, 147, 158
spatial arrangement
in chains or rings 62, 83, 86, 91
as factor in explanations 58, 60, 67, 76, 79, 86, 88, 89, 91, 124, 128, 153, 155
substances
compound 2, 4, 15, 22, 37, 39, 43–4, 51, 97–8, 105–6, 114, 119, 123, 128, 131, 133, 137, 152, 155
elementary 2, 4, 15, 22–3, 30, 32, 35, 37, 44, 47, 51, 113, 117, 121, 124–5, 131, 137, 152

imponderable or incorporeal 3–4, 18–19, 32, 37, 48, 108, 113
isolation and purification of 2, 4, 30, 36, 38, 41–2, 49, 84, 115, 121, 125, 138, 140
isomeric 60–1, 85–92, 128, 153, 155
mixed (mixtures) 2, 15, 28, 44, 49, 50, 91, 97–8, 105–6, 114, 123, 128, 133, 155
model 73–4, 78, 85–6
properties and dispositions of 9–10, 28, 30, 32, 34, 38, 48, 79, 103, 117, 124–5, 143
spectral signature of 124–5
ultimate (principles) 6, 8, 15, 18, 21, 23, 25, 28–9, 33, 36, 43, 47

theories
monolithic or overarching xi, 1, 9, 11, 16, 18, 37, 47, 157
incommensurability of 37, 47, 77, 128
standard model of 136–7, 150
truth of 11
truth 4–5, 10–11
type approach 57, 61, 65, 76–7, 80, 84, 87, 138

underdetermination xiii–xiv, 136–9, 152–4

valency 61–4, 66, 69, 79, 83, 93, 107

weight conservation 19–20, 28, 42